Heng Huat Chan
Theta functions, elliptic functions and π

Also of Interest

Algebra and Number Theory. A Selection of Highlights
Benjamin Fine, Anthony Gaglione, Anja Moldenhauer,
Gerhard Rosenberger, Dennis Spellman, 2017
ISBN 978-3-11-051584-8, e-ISBN (PDF) 978-3-11-051614-2,
e-ISBN (EPUB) 978-3-11-051626-5

Linear Algebra. A Course for Physicists and Engineers
Arak M. Mathai, Hans J. Haubold, 2017
ISBN 978-3-11-056235-4, e-ISBN (PDF) 978-3-11-056250-7,
e-ISBN (EPUB) 978-3-11-056259-0

Noncommutative Geometry. A Functorial Approach
Igor V. Nikolaev, 2017
ISBN 978-3-11-054317-9, e-ISBN (PDF) 978-3-11-054525-8,
e-ISBN (EPUB) 978-3-11-054348-3

*Abstract Algebra. Applications to Galois Theory, Algebraic
Geometry, Representation Theory and Cryptography*
Celine Carstensen-Opitz, Benjamin Fine, Anja Moldenhauer,
Gerhard Rosenberger, 2019
ISBN 978-3-11-060393-4, e-ISBN (PDF) 978-3-11-060399-6,
e-ISBN (EPUB) 978-3-11-060525-9

Commutative Algebra
Aron Simis, 2020
ISBN 978-3-11-061697-2, e-ISBN (PDF) 978-3-11-061698-9,
e-ISBN (EPUB) 978-3-11-061707-8

Philosophy of Mathematics
Thomas Bedürftig, Roman Murawski, 2018
ISBN 978-3-11-046830-4, e-ISBN (PDF) 978-3-11-046833-5,
e-ISBN (EPUB) 978-3-11-047077-2

Heng Huat Chan

Theta functions, elliptic functions and π

—

DE GRUYTER

Mathematics Subject Classification 2010
Primary: 33E05; Secondary: 14H99

Author
Prof. Heng Huat Chan
National University of Singapore
Department of Mathematics
10 Lower Kent Ridge Road
Block S17
Singapore 119076
Singapore
matchh@nus.edu.sg

ISBN 978-3-11-054071-0
e-ISBN (PDF) 978-3-11-054191-5
e-ISBN (EPUB) 978-3-11-054075-8

Library of Congress Control Number: 2020937803

Bibliographic information published by the Deutsche Nationalbibliothek
The Deutsche Nationalbibliothek lists this publication in the Deutsche Nationalbibliografie;
detailed bibliographic data are available on the Internet at http://dnb.dnb.de.

© 2020 Walter de Gruyter GmbH, Berlin/Boston
Cover image: sakkmesterke / iStock / Getty Images Plus
Typesetting: VTeX UAB, Lithuania
Printing and binding: CPI books GmbH, Leck

www.degruyter.com

To my wife Siew Lian Tan and our three children Si Min, Si Ya and Si En

Contents

Foreword

When one peruses the offerings of mathematics book publishers, one finds a large variety of text books at the undergraduate level, and also at the advanced level for upper-level graduate students and researchers. However, few books are "in between." *Theta functions, elliptic functions and π* falls "in between." This book, focusing on certain classical topics related to number theory, provides a stepping stone to both the past and future. The topics are chosen both for their elegance and their usefulness. The ubiquitous theta functions play the leading role. They form relationships to elliptic functions, sums of squares, partitions, hypergeometric functions, q-series, and infinite series representations for $1/\pi$. This book is valuable, because some of its topics do not appear in any of the courses taught by large, major universities. But it is even more valuable because it is inspirational. As you read it, you will exclaim, "What a beautiful theorem!" or "What an elegant proof!" or "What an interesting problem!" (and, indeed, there are many challenging exercises). Read with enjoyment!

University of Illinois at Urbana-Champaign   Bruce Berndt

https://doi.org/10.1515/9783110541915-201

Introduction

Theta functions and elliptic functions are usually discussed one after the other in classical books such as "*A Course of Modern Analysis*" by Whittaker and Watson [72]. The number π, on the other hand, appears in both popular and technical mathematics books. The three topics were rarely presented together in a book until the appearance of "*Pi and the AGM*" by Borwein and Borwein [13]. The Borweins' book is an excellent source of beautiful identities and algorithms involving theta functions and elliptic functions which are closely connected with the computations of π. There are generalizations of Ramanujan's series for $1/\pi$ [13, Chapter 5], elegant iterations to π motivated by the Gauss–Brent–Salamin algorithm [13, Chapter 2], and an interesting discussion of the Rogers–Ramanujan identities [13, Chapter 3] and many more that I highly recommend the interested reader to explore.

In 1996, I attempted to conduct a course for my graduate students at National Chung Cheng University (Taiwan) using the Borweins' book as a main reference. I then realized that it was almost impossible to teach based on the materials in the book as it requires students to have good understanding of topics like hypergeometric series, theta functions, elliptic functions and modular forms. From then on, I believed that a book serving as a bridge for advanced undergraduates to the materials in the Borweins' book is needed. I hope the present book will serve this purpose.

There are nine chapters in this book. The first chapter gives the proof of Jacobi's triple product identity (due to Andrews [2]) using Cauchy's generalization of the binomial theorem. Jacobi's triple product identity is an elegant identity that expresses a certain series of two variables in terms of three (hence the term "triple") infinite products. I also introduce the quintuple product identity (a formula expressing a certain series of two variables in terms of five infinite products) in the final section of this chapter. There are many proofs of the quintuple product identity and they are discussed in a comprehensive survey paper by Cooper [39]. Among these proofs, I choose to present the proof by Carlitz and Subbarao [20]. Their proof involves only the use of Jacobi's triple product identity and is an excellent example illustrating an application of this identity. An application of both Jacobi's triple product identity and the quintuple product identity will be given in Chapter 4.

Jacobi's triple product identity is an identity involving the two variables q and z. Specializing z leads to interesting identities involving only q. This brought up naturally the study of Jacobi's theta functions of one variable. I begin Chapter 2 with the definitions of three theta functions (due to Jacobi) involving only the vari-

https://doi.org/10.1515/9783110541915-202

able q. I do not assume Jacobi's triple product identity in this chapter but instead derive properties satisfied by these theta functions and show how to derive Jacobi's triple product identity using these identities. The approach in this chapter is similar to that in Chapter 2 of the Borweins' book [13, Chapter 2]. In the last section of this chapter, I introduce a function $\lambda(q)$ which will play an important role in this book.

It is natural to consider extending Jacobi's theta functions of one variable to two variables, which we thus defined in Chapter 3. It is in this chapter that I set $q = e^{\pi i \tau}$ and this will be used for the rest of the book. Using Jacobi's triple product identity, it is shown that these theta functions of two variables have infinite product representations. An immediate consequence is the derivation of an important identity of Jacobi, which plays a crucial role in Ramanujan's proofs of two of his congruences for the partition function $p(n)$.

In Chapter 4, I use the infinite product representation of Jacobi's theta function $\vartheta_1(u|\tau)$ and derive a series expansion for $\vartheta_1'(u|\tau)/\vartheta_1(u|\tau)$. I then introduce Lambert's series $L_{2j}(q^2)$ as a constant multiple of the coefficients of u^j in this series expansion. Using Jacobi's triple product identity and the quintuple product identity, I derive three important differential equations that are now known as Ramanujan's differential equations. I then present Ramanujan's proof of a well-known identity that expresses Dedekind's function $\Delta(\tau)$ in terms of $L_4(q^2)$ and $L_6(q^2)$. This is an identity that is usually derived using the theory of modular forms, but Ramanujan's derivation of this identity from the three differential equations is more elementary.

I introduce elliptic functions in Chapter 5. As in Chapter 2, I do not assume Jacobi's triple product identity. I first discuss the basic properties of elliptic functions and apply these properties to derive Jacobi's identity, which expresses $\vartheta_1'(0|\tau)$ in terms of $\vartheta_2(0|\tau)$, $\vartheta_3(0|\tau)$ and $\vartheta_4(0|\tau)$. To illustrate the importance of this identity, I present a third proof of Jacobi's triple product identity. This identity also plays an essential role in the derivation of a differential equation satisfied by one of Jacobi's elliptic functions introduced in Chapter 7.

Theta functions, Jacobi's elliptic functions and Weierstrass' $\wp$ function are usually discussed separately in most existing books. Few books discuss identities satisfied by these functions and the consequences arising from these identities. In Chapter 6 and Chapter 7, I introduce Weierstrass' $\wp$ function and two elliptic functions constructed from Jacobi's theta functions of two variables. I then derive identities satisfied by these three functions and establish relations between Lambert's series $L_{2j}(q^2)$, $\lambda(q)$ and $\vartheta_3(0|\tau)$. A transformation formula for Dedekind's function $\Delta(\tau)$ is derived as a consequence. This formula is usually proved using the theory of modular forms by considering $\Delta(\tau)$ as a weight 12 modular form on $SL_2(\mathbf{Z})$.

In Chapter 8, I use the identities derived in Chapter 7 to construct a differential equation satisfied by $\lambda(q)$ and $\vartheta_3^2(0|\tau)$. In the process of the derivation, I give a proof of Lagrange's four-square theorem. The differential equation satisfied by $\lambda(q)$ and $\vartheta_3^2(0|\tau)$ can be solved explicitly in terms of Gauss' hypergeometric function $_2F_1(a,b;c;z)$. I then derive a differential equation satisfied by $4\lambda(q)(1-\lambda(q))$ and $\vartheta_3^4(0|\tau)$ and show that these functions are related via Gauss' hypergeometric function $_3F_2(a,b,c;d,e;z)$. A consequence of this discussion yields a special case of Clausen's formula, which expresses $_3F_2(1/2,1/2,1/2;1,1;z)$ as the square of $_2F_1(1/4,1/4;1;z)$. A series of Ramanujan for $1/\pi$ is then derived from this identity and an identity that involves $\lambda(q)$ and $\lambda(q^2)$, which is established in Chapter 2.

I begin the discussion of Gauss' arithmetic–geometric mean in the last chapter. Using identities derived in Chapter 8, I derive an algorithm for computing π. This algorithm is due to Gauss (see [6, Chapter 7]) but was rediscovered by Brent [18] and Salamin [65]. The proof presented here, unlike that in the Borweins' book, does not involve elliptic integrals.

To summarize, the book can be divided into three parts, namely, Chapter 1 to Chapter 4, Chapter 5 to Chapter 7 and finally Chapter 8 and Chapter 9. I hope the reader will enjoy the identities explored in this book as much as I do and will be motivated to learn more about elliptic functions, theta functions, series for $1/\pi$, algorithms for computing π and the relations between these topics and modular forms.

National University of Singapore   Heng Huat Chan

Acknowledgments

This book would not have been written without the encouragement of my colleagues, friends and family.

In 2013, when I was on sabbatical leave at the University of Hong Kong, my host Kai Man Tsang encouraged me to present a series of lectures. Kai Man and his colleague Yuk-Kam Lau then suggested that I compiled my lecture notes into book form. I am grateful to both of them for this suggestion, which marks the beginning of this book project. The first draft of the book was typeset by Yingnan Wang and I thank him for his typesetting and for correcting several errors of the first version of this book.

In 2015, I conducted a short course at the East China Normal University based on the materials in this book. The second version of this book was done during this period. I am grateful to Zhi-Guo Liu for providing me an opportunity to do so.

The next revision of this book was done during my sabbatical leave at the University of Vienna in 2016. I thank my host Christian Krattenthaler and his colleague Michael J. Schlosser for their kind hospitality during my stay in Vienna.

This book was finally completed during my sabbatical leave at National Changhua University of Education (Taiwan) in 2020. I thank the university for providing an excellent working environment and my host Sen-Shan Huang, his wife Jen-Yin Chen and his colleague Kuo-Jye Chen for their warm hospitality.

There are many misprints and errors in various drafts of this book. First and foremost, I wish to mention that Shaun Cooper, who has read the drafts a few times and provided me with many invaluable criticisms and suggestions. I am deeply appreciative of his help. Other mathematicians such as Bruce C. Berndt, Gaurav Bhatnagar, Sen-Shan Huang, Yingnan Wang and LiuQuan Wang have also spent their precious time to read the drafts and improve the book. I am greatly indebted to all these mathematicians for their meticulous proof reading.

Next, I would like to thank George E. Andrews for his constant encouragement and for sharing his personal experience on his discovery of his proof of Jacobi's triple product identity.

My PhD advisor B. C. Berndt has given me enormous support and encouragement since 1991. I am extremely grateful to him for agreeing to write the foreword for this book.

The staff at De Gruyter have been very helpful. I wish to thank Chao Yang for handling the book proposal, Nadja Schedensack for her advice and the choice of the book cover and Ina Talandienė for her wonderful effort in editing the final version of my LaTeX files.

Finally, I thank my wife Siew Lian Tan for proof reading the final draft and for her love, support and encouragement all these years.

https://doi.org/10.1515/9783110541915-203

1 An introduction to Jacobi's triple product identity

In this chapter, we study Cauchy's generalization of the binomial theorem and derive two identities due to Euler. We then present Andrews' proof of Jacobi's triple product identity. We end the chapter with a proof, due to Carlitz and Subbarao, of the quintuple product identity using Jacobi's triple product identity.

1.1 Cauchy's generalization of the binomial theorem

The binomial theorem states that, if x is a complex number and n is a positive integer, then

$$(1 + x)^n = \sum_{j=0}^{n} \binom{n}{j} x^j, \tag{1.1}$$

where

$$\binom{n}{j} = \frac{n!}{j!(n-j)!}.$$

The numbers $\binom{n}{j}$ are called binomial coefficients.

Example 1.1. The following are instances of (1.1) for $n = 2, 3$ and 4:

$$(1 + x)^2 = 1 + 2x + x^2,$$
$$(1 + x)^3 = 1 + 3x + 3x^2 + x^3,$$

and

$$(1 + x)^4 = 1 + 4x + 6x^2 + 4x^3 + x^4.$$

The binomial theorem has an interesting generalization due to Cauchy [22, (18)]. Before we describe Cauchy's identity, we need a few definitions.

Definition 1.1. Let a and q be complex numbers. Set $(a; q)_0 = 1$. For any positive integer n, let

$$(a; q)_n = \prod_{k=0}^{n-1} (1 - aq^k) = (1 - a)(1 - aq) \cdots (1 - aq^{n-1}).$$

For $|q| < 1$, we define

$$(a; q)_\infty = \prod_{k=0}^{\infty} (1 - aq^k).$$

https://doi.org/10.1515/9783110541915-001

Definition 1.2. Following the notation in "Special Functions" by Andrews, Askey and Roy [3, (10.0.8)], let

$$0!_q = 1$$

and

$$n!_q = 1 \times (1+q) \times (1+q+q^2) \times \cdots \times (1+q+q^2+\cdots+q^{n-1}) \quad \text{when } n \in \mathbf{Z}^+. \quad (1.2)$$

Observe that when $q = 1$, $n!_q$ reduces to $n!$.

Summing the geometric series in each term of the right-hand side of (1.2), we find that

$$n!_q = \frac{(q;q)_n}{(1-q)^n}.$$

Definition 1.3. Let j be an integer satisfying $0 \le j \le n$. We define the q-analogue of the binomial coefficient as

$$\begin{bmatrix} n \\ j \end{bmatrix}_q = \frac{n!_q}{j!_q(n-j)!_q} = \frac{(q;q)_n}{(q;q)_j(q;q)_{n-j}}. \quad (1.3)$$

We observe that

$$\lim_{q \to 1} \begin{bmatrix} n \\ j \end{bmatrix}_q = \binom{n}{j}. \quad (1.4)$$

With the notation given in (1.3), we now state the following theorem of Cauchy.

Theorem 1.4. *Let n be a positive integer, z and q be complex numbers. Then*

$$(-z;q)_n = \sum_{j=0}^{n} \begin{bmatrix} n \\ j \end{bmatrix}_q q^{j(j-1)/2} z^j. \quad (1.5)$$

Example 1.2. The following are examples of (1.5) for $n = 2, 3, 4$:

$$(1+z)(1+zq) = \begin{bmatrix} 2 \\ 0 \end{bmatrix}_q + \begin{bmatrix} 2 \\ 1 \end{bmatrix}_q z + \begin{bmatrix} 2 \\ 2 \end{bmatrix}_q qz^2$$

$$= 1 + (1+q)z + qz^2,$$

$$(1+z)(1+zq)(1+zq^2) = \begin{bmatrix} 3 \\ 0 \end{bmatrix}_q + \begin{bmatrix} 3 \\ 1 \end{bmatrix}_q z + \begin{bmatrix} 3 \\ 2 \end{bmatrix}_q qz^2 + \begin{bmatrix} 3 \\ 3 \end{bmatrix}_q q^3 z^3$$

$$= 1 + (1+q+q^2)z + (1+q+q^2)qz^2 + q^3 z^3,$$

and

$$(1+z)(1+zq)(1+zq^2)(1+zq^3)$$

$$= \begin{bmatrix} 4 \\ 0 \end{bmatrix}_q + \begin{bmatrix} 4 \\ 1 \end{bmatrix}_q z + \begin{bmatrix} 4 \\ 2 \end{bmatrix}_q qz^2 + \begin{bmatrix} 4 \\ 3 \end{bmatrix}_q q^3 z^3 + \begin{bmatrix} 4 \\ 4 \end{bmatrix}_q q^6 z^3$$

$$= 1 + (1 + q + q^2 + q^3)z + (1 + q^2)(1 + q + q^2)qz^2$$
$$+ (1 + q + q^2 + q^3)q^3 z^3 + q^6 z^4.$$

We observe that when $q = 1$ and $z = x$, the identities in Example 1.2 reduce to the identities in Example 1.1. It is immediate from (1.4) that (1.5) reduces to (1.1) when $q = 1$ and $z = x$.

We now present the proof of Theorem 1.4.

Proof. Let

$$f(z) = (-z; q)_n. \tag{1.6}$$

Since $f(z)$ is a polynomial in z of degree n, we may let

$$f(z) = \sum_{j=0}^{n} Q_j(q)z^j, \tag{1.7}$$

where $Q_0(q) = 1$ and $Q_j(q)$ is a polynomial in q for $1 \le j \le n$.

Multiplying $f(zq)$ by $(1 + z)$ and using (1.6), we obtain the relation

$$(1 + z)f(zq) = (1 + z)(1 + zq) \cdots (1 + zq^{n-1})(1 + zq^n)$$
$$= f(z)(1 + zq^n). \tag{1.8}$$

Using (1.7) and (1.8), we deduce that

$$(1 + z) \sum_{j=0}^{n} q^j Q_j(q)z^j = \sum_{j=0}^{n} Q_j(q)z^j(1 + zq^n). \tag{1.9}$$

Comparing the coefficients of z^j on both sides of (1.9), we conclude that

$$q^j Q_j(q) + q^{j-1}Q_{j-1}(q) = Q_j(q) + q^n Q_{j-1}(q),$$

which simplifies as

$$(1 - q^j)Q_j(q) = q^{j-1}(1 - q^{n-j+1})Q_{j-1}(q).$$

The last identity implies that

$$Q_j(q) = q^{j-1}\frac{(1 - q^{n-j+1})}{(1 - q^j)}Q_{j-1}(q). \tag{1.10}$$

Iterating (1.10), we deduce that

$$Q_j(q) = q^{(j-1)+(j-2)+\cdots2+1}\frac{(1 - q^{n-j+1})(1 - q^{n-j+2})\cdots(1 - q^n)}{(1 - q^j)(1 - q^{j-1})\cdots(1 - q^2)(1 - q)}Q_0(q)$$

$$= q^{j(j-1)/2}\begin{bmatrix} n \\ j \end{bmatrix}_q,$$

where the last equality holds since $Q_0(q) = 1$. This completes the proof of (1.5). $\square$

The proof of Theorem 1.4 we presented here can be found in the article by Alexanderson and Polya [1].

1.2 Jacobi's triple product identity

One of the most elegant two-variable identities connecting a series with an infinite product is Jacobi's triple product identity. This is stated as follows.

Theorem 1.5 (Jacobi's triple product identity). *Let z be a non-zero complex number, and let q be a complex number such that $|q| < 1$. Then*

$$\sum_{j=-\infty}^{\infty} q^{j^2}z^j = (-zq; q^2)_\infty(-z^{-1}q; q^2)_\infty(q^2; q^2)_\infty. \tag{1.11}$$

There are several proofs of (1.11) in the literature. One of the proofs was given by Jacobi [51, Section 64] using q-series and this was reproduced in the book by Gasper and Rahman [44, p. 12]. Another proof of (1.11), which involves the clever use of $\sqrt{-1}$, can be found in the book by Hardy and Wright [48, Section 19.8]. Two other proofs of (1.11) using theory of elliptic functions and properties of Jacobi's theta functions are discussed in the book by Whittaker and Watson [72]. For more details, see [72, p. 469–473] and [72, p. 490]. Identity (1.11) can also be established in a combinatorial way through the "boson–fermion correspondence", an idea due to Borcherds. For more details of this approach, see Hei-Chi Chan's book [23, Chapter 5] and the recent book of Shun-Jen Cheng and Weiqiang Wang [38, Section A4]. It is well known that (1.11) is a special case of Macdonald's identities [59].

A detailed discussion of this approach can be found in Carter's book [21, Chapter 20].

In this book, we will discuss three proofs of Jacobi's triple product identity. Our first proof, which will be presented in this chapter, is due to Andrews [2]. The second proof and third proof will be presented in Chapter 2 and Chapter 5, respectively.

Andrews' proof of (1.11) requires only Theorem 1.4 established in Section 1.1. Before discussing the proof in Section 1.4, let us consider a few interesting specializations of (1.11).

Example 1.3. Let $z = 1$ in (1.11). We immediately obtain a product representation of the generating function for squares, namely,

$$\sum_{j=-\infty}^{\infty} q^{j^2} = (-q; q^2)_\infty^2 (q^2; q^2)_\infty.$$

We remark here that the squares $j^2 = 1, 4$ and 9 for $j = 1, 2$ and 3, respectively, can be viewed as the number of dots in Figure 1.1.

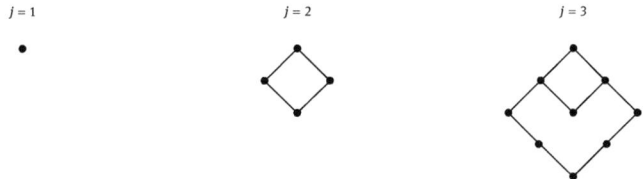

$j = 1$ $j = 2$ $j = 3$

Figure 1.1: Visualization of squares.

Example 1.4. Let $z = q$ in (1.11) and replace q by $q^{1/2}$ to deduce that

$$\sum_{j=-\infty}^{\infty} q^{j(j+1)/2} = 2(-q; q)_\infty^2 (q; q)_\infty.$$

Simplifying the above identity, we find that

$$\sum_{j=0}^{\infty} q^{j(j+1)/2} = (-q; q)_\infty^2 (q; q)_\infty.$$

The integers $j(j + 1)/2$ are the triangular numbers. For $j = 1, 2$ and 3, they are $1, 3$ and 6, respectively and these are the number of dots in Figure 1.2.

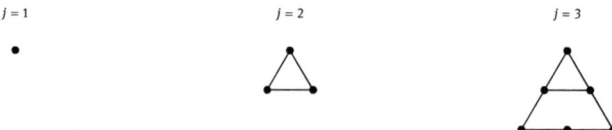

Figure 1.2: Visualization of triangular numbers.

Example 1.5. Let q be replaced by $q^{3/2}$ and $z = q^{-1/2}$. Using (1.11), we deduce that

$$\sum_{j=-\infty}^{\infty} q^{j(3j-1)/2} = (-q; q^3)_\infty (-q^2; q^3)_\infty (q^3; q^3)_\infty.$$

The integers of the form $\omega(j) = j(3j - 1)/2$, with $j \geq 1$, are known as pentagonal numbers. For $j = 1, 2$ and 3, $w(j) = 1, 5$ and 12, respectively. These correspond, respectively, to the number of dots in Figure 1.3.

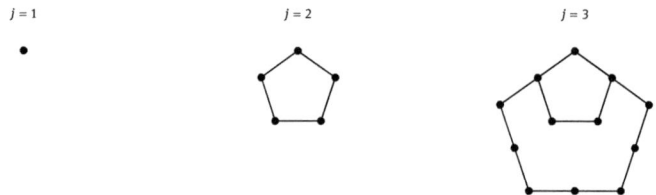

Figure 1.3: Visualization of pentagonal numbers.

Example 1.6. There is an important identity associated with the pentagonal numbers due to Euler. If we replace q by $q^{3/2}$ and let $z = -q^{-1/2}$, then we obtain the identity

$$\sum_{j=-\infty}^{\infty} (-1)^j q^{j(3j-1)/2} = (q; q^3)_\infty (q^2; q^3)_\infty (q^3; q^3)_\infty = (q; q)_\infty. \tag{1.12}$$

1.3 Two identities of Euler

The materials in this section are based on the article by Chan, Chan and Cooper [29].

Andrews' proof of (1.11) mentioned in Section 1.2 requires two of Euler's identities (1.14) and (1.16), which follow from Theorem 1.4.

We will need the following lemma known as Tannery's theorem (see Bromwich's book [19, § 49]).

Lemma 1.6. *For each positive integer n, let*

$$\sum_{j=0}^{p_n} v_j(n)$$

be a finite sum where $p_n \to \infty$ as $n \to \infty$.

If for each j, the limit $\lim_{n\to\infty} v_j(n)$ exists and there is a convergent series

$$\sum_{j=0}^{\infty} M_j$$

of non-negative real numbers such that, for all integers $j \geq 0$, $n \geq 1$,

$$|v_j(n)| \leq M_j,$$

then

$$\lim_{n\to\infty} \sum_{j=0}^{p_n} v_j(n) = \sum_{j=0}^{\infty} \lim_{n\to\infty} v_j(n).$$

We first rewrite (1.5) as

$$\frac{(1+z)(1+zq)\cdots(1+zq^{n-1})}{(q;q)_n} = \sum_{j=0}^{n} \frac{q^{j(j-1)/2}}{(q;q)_j(q;q)_{n-j}} z^j. \tag{1.13}$$

Suppose $0 < q < 1$. Letting $n \to \infty$ in (1.13) and applying Lemma 1.6 with

$$M_j = \frac{1}{(q;q)_\infty^2} q^{j(j-1)/2} |z|^j,$$

we deduce the following identity of Euler for $0 < q < 1$ and extend the identity to include complex numbers q with $|q| < 1$ by analytic continuation.

Theorem 1.7. *Let z and q be complex numbers and $|q| < 1$. Then*

$$(-z;q)_\infty = \sum_{j=0}^{\infty} \frac{q^{j(j-1)/2}}{(q;q)_j} z^j. \tag{1.14}$$

Next, set $z = -1$ in (1.5) to deduce that, for $n \geq 1$,

$$0 = \sum_{j=0}^{n} \frac{1}{(q;q)_{n-j}} \cdot \frac{(-1)^j q^{j(j-1)/2}}{(q;q)_j}. \tag{1.15}$$

We observe that the right-hand side of (1.15) is the coefficient of z^n in the expansion of $A(z)B(z)$ for each positive integer n, where

$$A(z) = \sum_{j=0}^{\infty} \frac{z^j}{(q;q)_j} \quad \text{and} \quad B(z) = \sum_{j=0}^{\infty} \frac{(-1)^j q^{j(j-1)/2}}{(q;q)_j} z^j.$$

Hence, by (1.15), we conclude that

$$A(z)B(z) = 1.$$

Using (1.14) with z replaced by $-z$, we arrive at the following identity of Euler.

Theorem 1.8. *For complex numbers z and q with* $|q| < 1$,

$$\sum_{j=0}^{\infty} \frac{z^j}{(q;q)_j} = \frac{1}{(z;q)_{\infty}}. \tag{1.16}$$

Identities (1.14) and (1.16) can be established by purely combinatorial considerations. For more details, see [48, Section 19.5].

Let

$$E_q(z) = \sum_{j=0}^{\infty} \frac{q^{j(j-1)/2}}{(q;q)_j} z^j$$

and

$$e_q(z) = \sum_{j=0}^{\infty} \frac{z^j}{(q;q)_j}.$$

Since

$$\lim_{q \to 1^-} E_q(z(1-q)) = e^z$$

and

$$\lim_{q \to 1^-} e_q(z(1-q)) = e^z,$$

the functions $E_q(z)$ and $e_q(z)$ can both be viewed as q-analogues of the exponential function. From the product representations (1.14) and (1.16) of these two functions, we observe that

$$e_q(-z)E_q(z) = e_q(z)E_q(-z) = 1,$$

which are both analogues of $e^z e^{-z} = 1$.

1.4 Andrews' proof of Jacobi's triple product identity

We are now ready to present Andrews' proof of (1.11).

Proof of (1.11). Replacing q by q^2 in (1.14), we deduce that

$$(-z; q^2)_\infty = \sum_{\ell=0}^{\infty} \frac{q^{\ell(\ell-1)}}{(q^2; q^2)_\ell} z^\ell. \tag{1.17}$$

Replacing z by zq in (1.17), we find that

$$(-zq; q^2)_\infty = \sum_{\ell=0}^{\infty} \frac{q^{\ell^2}}{(q^2; q^2)_\ell} z^\ell. \tag{1.18}$$

Observing that

$$\frac{1}{(q^2; q^2)_\ell} = \frac{(q^{2\ell+2}; q^2)_\infty}{(q^2; q^2)_\infty},$$

we rewrite (1.18) as

$$\begin{aligned}
(-zq; q^2)_\infty &= \sum_{\ell=0}^{\infty} \frac{(q^{2\ell+2}; q^2)_\infty}{(q^2; q^2)_\infty} q^{\ell^2} z^\ell \\
&= \frac{1}{(q^2; q^2)_\infty} \sum_{\ell=0}^{\infty} (q^{2\ell+2}; q^2)_\infty q^{\ell^2} z^\ell \\
&= \frac{1}{(q^2; q^2)_\infty} \sum_{\ell=-\infty}^{\infty} (q^{2\ell+2}; q^2)_\infty q^{\ell^2} z^\ell,
\end{aligned}$$

where in the last equality we have used the fact that, for $\ell < 0$,

$$(q^{2\ell+2}; q^2)_\infty = 0.$$

This implies that

$$(-zq; q^2)_\infty (q^2; q^2)_\infty = \sum_{\ell=-\infty}^{\infty} (q^{2\ell+2}; q^2)_\infty q^{\ell^2} z^\ell. \tag{1.19}$$

Next, using (1.17) with $z = -q^{2\ell+2}$, we find that

$$(q^{2\ell+2}; q^2)_\infty = \sum_{\nu=0}^{\infty} (-1)^\nu \frac{q^{\nu^2+\nu+2\nu\ell}}{(q^2; q^2)_\nu}.$$

Substituting this expression into the right-hand side of (1.19), we deduce that

$$(-zq; q^2)_\infty (q^2; q^2)_\infty = \sum_{\ell=-\infty}^{\infty} q^{\ell^2} z^\ell \sum_{\nu=0}^{\infty} (-1)^\nu \frac{q^{\nu^2+\nu+2\nu\ell}}{(q^2; q^2)_\nu}$$

$$= \sum_{v=0}^{\infty} \sum_{\ell=-\infty}^{\infty} \frac{(-q)^v q^{(\ell+v)^2}}{(q^2;q^2)_v} z^\ell$$

$$= \sum_{v=0}^{\infty} \frac{(-q/z)^v}{(q^2;q^2)_v} \sum_{\ell=-\infty}^{\infty} q^{(\ell+v)^2} z^{\ell+v}$$

$$= \sum_{v=0}^{\infty} \frac{(-q/z)^v}{(q^2;q^2)_v} \sum_{\ell=-\infty}^{\infty} q^{\ell^2} z^\ell$$

$$= \frac{1}{(-z^{-1}q;q^2)_\infty} \sum_{\ell=-\infty}^{\infty} q^{\ell^2} z^\ell, \tag{1.20}$$

where we have used the fact that, for all $v \in \mathbf{Z}$,

$$\sum_{\ell=-\infty}^{\infty} q^{(\ell+v)^2} z^{\ell+v} = \sum_{\ell=-\infty}^{\infty} q^{\ell^2} z^\ell$$

in the second last equality, and

$$\sum_{v=0}^{\infty} \frac{(-q/z)^v}{(q^2;q^2)_v} = \frac{1}{(-z^{-1}q;q^2)_\infty} \tag{1.21}$$

in the last equality. Identity (1.21) follows from (1.16) with q replaced by q^2, followed by replacing z by $-q/z$. Combining (1.19) and (1.20), we conclude the proof of Jacobi's triple product identity. ☐

1.5 Jacobi's triple product identity and the proof of the quintuple product identity

The Jacobi triple product identity may be viewed as an identity expressing the product of three infinite products in terms of an infinite series. This identity has a companion that involves five infinite products. It is known as the quintuple product identity and is given as follows.

Theorem 1.9 (Quintuple product identity). *Let $q, t \in \mathbf{C}$ with $|q| < 1$. Then*

$$(q^2;q^2)_\infty (tq^2;q^2)_\infty (t^{-1};q^2)_\infty (t^2q^2;q^4)_\infty (t^{-2}q^2;q^4)_\infty = \sum_{j=-\infty}^{\infty} q^{3j^2+j}(t^{3j} - t^{-3j-1}). \tag{1.22}$$

According to a recent finding of Liu, the quintuple product identity first appeared in a paper by Kiepert [54] in 1879. There are many proofs of (1.22). For an excellent discussion of these proofs, see the article by Cooper [39]. In this section, we present a proof of (1.22) due to Carlitz and Subbarao [20]. A reason for following their proof is that it requires only the knowledge of Jacobi's triple product identity (1.11).

Proof of Theorem 1.9. Let

$$A(q,t) = (q^4;q^4)_\infty (t^2q^2;q^4)_\infty (t^{-2}q^2;q^4)_\infty (q^2;q^2)_\infty (tq^2;q^2)_\infty (t^{-1};q^2)_\infty.$$

Using Jacobi's triple product identity (1.11), we find that

$$(q^4;q^4)_\infty (t^2q^2;q^4)_\infty (t^{-2}q^2;q^4)_\infty = \sum_{k=-\infty}^{\infty} (-1)^k q^{2k^2} t^{2k}$$

and

$$(q^2;q^2)_\infty (tq^2;q^2)_\infty (t^{-1};q^2)_\infty = \sum_{j=-\infty}^{\infty} (-1)^j q^{j^2} t^j q^j.$$

This implies that

$$A(q,t) = \sum_{k=-\infty}^{\infty} \sum_{j=-\infty}^{\infty} (-1)^k q^{2k^2} (-1)^j q^{j^2+j} t^{2k+j}$$

$$= \sum_{k=-\infty}^{\infty} \sum_{n=-\infty}^{\infty} (-1)^k q^{2k^2} t^{2k} (-1)^{n-2k} t^{n-2k} q^{(n-2k)^2+(n-2k)},$$

where we have replaced j by $n - 2k$ in the last equality. This implies that

$$A(q,t) = \sum_{n=-\infty}^{\infty} (-1)^n t^n q^{n^2+n} \sum_{k=-\infty}^{\infty} (-1)^k q^{6k^2-4nk-2k}$$

$$= \sum_{n=-\infty}^{\infty} (-1)^n t^n q^{n^2+n} \sum_{k=-\infty}^{\infty} (-1)^k q^{6k^2-2k(2n+1)}.$$

We now write

$$A(q,t) = A_0(q,t) + A_1(q,t) + A_2(q,t),$$

where

$$A_\nu(q,t) = \sum_{\substack{n=-\infty \\ 2n+1\equiv \nu \pmod 3}}^{\infty} (-1)^n q^{n^2+n} t^n \sum_{k=-\infty}^{\infty} (-1)^k q^{6k^2-2k(2n+1)}. \qquad (1.23)$$

We will first compute $A_0(q,t)$. If $2n + 1 \equiv 0 \pmod{3}$, then $3\ell = 2n + 1$ for some integer ℓ. Note that

$$\sum_{k=-\infty}^{\infty} (-1)^k q^{6k^2-6\ell k} = \sum_{k=-\infty}^{\infty} (-1)^{k+\ell} q^{6(k+\ell)^2-6\ell(k+\ell)}$$

$$= \sum_{k=-\infty}^{\infty} (-1)^{k+\ell} q^{6k^2+12k\ell-6\ell k}$$

$$= (-1)^{\ell} \sum_{k=-\infty}^{\infty} (-1)^k q^{6k^2-6\ell k}.$$

Note that ℓ is odd because $3\ell = 2n + 1$. Hence,

$$\sum_{k=-\infty}^{\infty} (-1)^k q^{6k^2-6\ell k} = 0.$$

We next find an expression for $A_1(q,t)$. Note that $2n + 1 \equiv 1 \pmod{3}$ implies that n is divisible by 3.

Therefore, by writing $n = 3s$, we find that

$$A_1(q,t) = \sum_{s=-\infty}^{\infty} (-1)^s q^{(3s)^2+3s} t^{3s} \sum_{k=-\infty}^{\infty} (-1)^k q^{6k^2-(12s+2)k}$$

$$= \sum_{s=-\infty}^{\infty} \sum_{k=-\infty}^{\infty} (-1)^{s+k} q^{9s^2+3s} t^{3s} q^{6k^2-12sk-2k}.$$

Next, let $k = m + s$. Then

$$A_1(q,t) = \sum_{m=-\infty}^{\infty} \sum_{s=-\infty}^{\infty} (-1)^m q^{9s^2+3s} q^{6(m+s)^2-12(m+s)s-2(m+s)} t^{3s}$$

$$= \sum_{s=-\infty}^{\infty} t^{3s} q^{3s^2+s} \sum_{m=-\infty}^{\infty} (-1)^m q^{6m^2-2m}$$

$$= \sum_{s=-\infty}^{\infty} t^{3s} q^{3s^2+s} (q^4; q^4)_{\infty},$$

where we have used (1.12), with q replaced by q^4, in the last step.

In a similar way, we can show that

$$A_2(q,t) = -(q^4; q^4)_{\infty} \sum_{s=-\infty}^{\infty} t^{-3s-1} q^{3s^2+s}.$$

Collecting all the identities for $A_\nu(q,t)$ for $\nu = 0, 1, 2$, we complete the proof of (1.22). $\qquad\square$

Exercises for Chapter 1

1. Let n and k be positive integers with $k < n$.

    (a) Show that

    $$\begin{bmatrix} n \\ k \end{bmatrix}_q = q^k \begin{bmatrix} n-1 \\ k \end{bmatrix}_q + \begin{bmatrix} n-1 \\ k-1 \end{bmatrix}_q$$

    and

    $$\begin{bmatrix} n \\ k \end{bmatrix}_q = \begin{bmatrix} n-1 \\ k \end{bmatrix}_q + q^{n-k} \begin{bmatrix} n-1 \\ k-1 \end{bmatrix}_q.$$

    (b) Deduce that $\begin{bmatrix} n \\ k \end{bmatrix}_q$ is a polynomial in q of degree $k(n-k)$.

2. Show that, if $|q| < 1$, then

    $$(q;q^5)_\infty (q^4;q^5)_\infty (q^5;q^5)_\infty = \sum_{k=-\infty}^{\infty} (-1)^k q^{k(5k+3)/2}$$

    and

    $$(q^2;q^5)_\infty (q^3;q^5)_\infty (q^5;q^5)_\infty = \sum_{k=-\infty}^{\infty} (-1)^k q^{k(5k+1)/2}.$$

3. Let $a, z, q \in \mathbf{C}$ with $|q| < 1$.

    (a) Use (1.5), (1.14) and (1.16) to show that

    $$\frac{(az;q)_\infty}{(z;q)_\infty} = \sum_{j=0}^{\infty} \frac{(a;q)_j}{(q;q)_j} z^j. \tag{1.24}$$

    (b) Identity (1.24) is often proved using ideas similar to the proof of Theorem 1.4. Assuming (1.24), derive (1.14) and (1.16).

4. Complete the following exercises which lead to a proof of an equivalent form of (1.11).

    (a) Use (1.5) to show that

    $$(z;q)_{2n} = \sum_{k=-n}^{n} \begin{bmatrix} 2n \\ n+k \end{bmatrix}_q (-1)^{k+n} q^{(k+n)(k+n-1)/2} z^{k+n}. \tag{1.25}$$

    (b) Use (1.25) to show that

    $$(z;q)_n (z^{-1}q;q)_n = (q;q)_{2n} \sum_{k=-n}^{n} \frac{(-1)^k q^{k(k-1)/2} z^k}{(q;q)_{n+k}(q;q)_{n-k}}.$$

(c) Deduce (1.11) using Tannery's theorem.

5. Show that, if $A_2(q, t)$ is given by (1.23), then

$$A_2(q,t) = -(q^4;q^4)_\infty \sum_{s=-\infty}^{\infty} t^{-3s-1} q^{3s^2+s}.$$

2 Jacobi's theta functions of one variable and the triple product identity

In this chapter, we introduce three of Jacobi's theta functions of one variable. We derive several fundamental identities satisfied by these functions and give another proof of Jacobi's triple product identity (1.11) presented by Borwein and Borwein in [13, Chapter 3]. We will also introduce $\lambda(q)$, a fourth power of a quotient of Jacobi's theta functions, and derive an identity satisfied by $\lambda(q)$ and $\lambda(q^2)$.

2.1 Jacobi's theta functions of one variable

We have encountered the generating functions of squares and triangular numbers in Chapter 1. We now relate them to classical theta functions of one variable, first studied extensively by Jacobi.

Definition 2.1. Let $q \in \mathbf{C}$ with $|q| < 1$. The *Jacobi theta functions of one variable* are defined by

$$\Theta_2(q) = \sum_{j=-\infty}^{\infty} q^{(j+1/2)^2},$$

$$\Theta_3(q) = \sum_{j=-\infty}^{\infty} q^{j^2},$$

and

$$\Theta_4(q) = \sum_{j=-\infty}^{\infty} (-1)^j q^{j^2}.$$

We observe that $\Theta_3(q)$ is the function discussed in Example 1.3. Rewriting $\Theta_2(q)$ as

$$\Theta_2(q) = q^{1/4} \sum_{j=-\infty}^{\infty} q^{j^2+j},$$

we find that $q^{-1/8}\Theta_2(q^{1/2})$ is the function discussed in Example 1.4. We have seen that if the summand of the series in Example 1.5 is "weighted" by $(-1)^j$, then we obtain Euler's identity (1.12). If we multiply the summand of the series in Example 1.3 by $(-1)^j$, we obtain $\Theta_4(q)$. The reader might notice that $\Theta_1(q)$ is not defined. The reason for its absence will be given in Chapter 3.

https://doi.org/10.1515/9783110541915-002

2.2 Identities associated with Jacobi's theta functions of one variable

There are many identities involving Jacobi's theta functions of one variable. Some of the most important identities will be discussed in this section and in Section 2.4.

Theorem 2.2. *For any complex number q with $|q| < 1$,*

$$\Theta_4^2(q^2) = \Theta_3(q)\Theta_4(q). \tag{2.1}$$

It is an exercise to show that Theorem 2.2 follows from (1.11). But since we aim to provide another proof of (1.11), we need to derive (2.1) without assuming (1.11).

Before proving (2.1), we need several identities satisfied by $\Theta_2(q)$, $\Theta_3(q)$ and $\Theta_4(q)$.

Lemma 2.3. *For complex number q with $|q| < 1$,*

$$\Theta_3^2(q) + \Theta_4^2(q) = 2\Theta_3^2(q^2). \tag{2.2}$$

Proof. Let

$$L = \{m + ni | m, n \in \mathbf{Z}\},$$

where $i = \sqrt{-1}$. Note that if $\ell = m + in \in L$ and

$$N(\ell) = m^2 + n^2,$$

then

$$\Theta_3^2(q) = \sum_{m,n=-\infty}^{\infty} q^{m^2+n^2} = \sum_{\ell \in L} q^{N(\ell)}.$$

We call $N(\ell)$ the norm of ℓ.

To prove (2.2), we first note that

$$\Theta_3^2(q) + \Theta_4^2(q) = \sum_{m,n=-\infty}^{\infty} (1 + (-1)^{m+n})q^{m^2+n^2}.$$

Since $1 + (-1)^{m+n} = 0$ if and only if $m \not\equiv n \pmod{2}$, we deduce that

$$\Theta_3^2(q) + \Theta_4^2(q) = 2 \sum_{\ell^* \in L^*} q^{N(\ell^*)},$$

where

$$L^* = \{m + ni | m \equiv n \ (\text{mod } 2)\}.$$

Let

$$L^\dagger = \{m(1 + i) + n(1 - i) | m, n \in \mathbf{Z}\}. \tag{2.3}$$

We claim that

$$L^* = L^\dagger. \tag{2.4}$$

To see this, we observe that if $\ell^\dagger \in L^\dagger$, then

$$\ell^\dagger = m(1 + i) + n(1 - i) = (m + n) + (m - n)i.$$

Since

$$m + n \equiv m - n \ (\text{mod } 2),$$

we deduce that $\ell^\dagger \in L^*$. Conversely, if $\ell^* \in L^*$, then

$$\ell^* = m + ni, \quad \text{with } m \equiv n \ (\text{mod } 2).$$

This implies that m and n have the same parity and, therefore, the numbers

$$r = \frac{m + n}{2}$$

and

$$s = \frac{m - n}{2}$$

are integers. Rewriting ℓ^* in terms of r and s, we find that

$$\ell^* = m + ni = r + s + (r - s)i = r(1 + i) + s(1 - i),$$

which implies that $\ell^* \in L^\dagger$. This completes the proof of (2.4).

Using (2.4), we conclude that

$$\begin{aligned}
\Theta_3^2(q) + \Theta_4^2(q) &= 2 \sum_{\ell^* \in L^*} q^{N(\ell^*)} \\
&= 2 \sum_{m,n \in \mathbf{Z}} q^{N((m+n)+(m-n)i)} \\
&= 2 \sum_{m,n=-\infty}^{\infty} q^{2(m^2+n^2)} \\
&= 2\Theta_3^2(q^2),
\end{aligned} \tag{2.5}$$

which completes the proof of (2.2). $\qquad\qquad\qquad\qquad\qquad\qquad$ □

For a proof of (2.2) without the consideration of lattices, namely the use of (2.4), see [13, p. 34].

We now complete the proof of Theorem 2.2.

Proof of Theorem 2.2. We begin with the identity

$$AB = \frac{1}{2}((A + B)^2 - A^2 - B^2). \tag{2.6}$$

Let $A = \Theta_3(q)$ and $B = \Theta_4(q)$ in (2.6). To prove (2.1), it suffices to simplify $(A + B)^2$ and $A^2 + B^2$. The identity

$$A^2 + B^2 = 2\Theta_3^2(q^2) \tag{2.7}$$

follows from (2.2).

Next, from the formula

$$\Theta_3(q) + \Theta_4(q) = 2 \sum_{j=-\infty}^{\infty} q^{4j^2}$$
$$= 2\Theta_3(q^4),$$

we conclude that

$$(A + B)^2 = (\Theta_3(q) + \Theta_4(q))^2$$
$$= 4\Theta_3^2(q^4). \tag{2.8}$$

Substituting (2.8) and (2.7) in (2.6), we deduce that

$$\Theta_3(q)\Theta_4(q) = \frac{1}{2}(4\Theta_3^2(q^4) - 2\Theta_3^2(q^2))$$
$$= 2\Theta_3^2(q^4) - \Theta_3^2(q^2). \tag{2.9}$$

From (2.2) with q replaced by q^2, we deduce that

$$2\Theta_3^2(q^4) = \Theta_3^2(q^2) + \Theta_4^2(q^2). \tag{2.10}$$

Substituting (2.10) into (2.9), we complete our proof of (2.1). □

2.3 Jacobi's triple product identity revisited

We now give the Borweins' proof of Jacobi's triple product identity using identities from the previous section. Let

$$F(z, q) = (-zq; q^2)_\infty (-z^{-1}q; q^2)_\infty.$$

Note that

$$F(z, q) = F(z^{-1}, q) \tag{2.11}$$

and

$$
\begin{aligned}
F(zq^2, q) &= \prod_{k=1}^{\infty} (1 + zq^{2k+1})(1 + z^{-1}q^{2k-3}) \\
&= \frac{1 + z^{-1}q^{-1}}{1 + zq} F(z, q) \\
&= z^{-1}q^{-1} F(z, q).
\end{aligned} \tag{2.12}
$$

Suppose that

$$F(z, q) = \sum_{j=-\infty}^{\infty} c_j(q) z^j. \tag{2.13}$$

From (2.13) and (2.12), we deduce that

$$\sum_{j=-\infty}^{\infty} c_j(q) z^j q^{2j} = \sum_{j=-\infty}^{\infty} c_j(q) z^{j-1} q^{-1},$$

which implies, by comparing coefficients of z^{j-1}, that

$$c_{j-1}(q) q^{2j-2} = c_j(q) q^{-1}$$

or

$$c_j(q) = q^{2j-1} c_{j-1}(q). \tag{2.14}$$

Let $j \geq 0$. By iterating (2.14), we deduce that

$$c_j(q) = q^{2j-1} q^{2j-3} \cdots q c_0(q) = q^{j^2} c_0(q).$$

By (2.11), we find that

$$c_j(q) = c_{-j}(q)$$

and this shows that

$$F(z, q) = c_0(q) \sum_{j=-\infty}^{\infty} q^{j^2} z^j.$$

It remains to determine $c_0(q)$. Note that we have

$$(-zq; q^2)_\infty (-z^{-1}q; q^2)_\infty = c_0(q) \sum_{j=-\infty}^{\infty} q^{j^2} z^j. \tag{2.15}$$

Substituting $z = 1$ and $z = -1$ in (2.15), we deduce that

$$(-q; q^2)_\infty^2 = c_0(q)\Theta_3(q) \tag{2.16}$$

and

$$(q; q^2)_\infty^2 = c_0(q)\Theta_4(q), \tag{2.17}$$

respectively. Multiplying (2.16) and (2.17), we find that

$$\begin{aligned}
\Theta_3(q)\Theta_4(q) &= c_0^{-2}(q)(-q; q^2)_\infty^2 (q; q^2)_\infty^2 \\
&= c_0^{-2}(q)(q^2; q^4)_\infty^2.
\end{aligned} \tag{2.18}$$

Next, by (2.1) and (2.17) (with q replaced by q^2), we find that

$$\begin{aligned}
\Theta_3(q)\Theta_4(q) &= \Theta_4^2(q^2) \\
&= c_0^{-2}(q^2)(q^2; q^4)_\infty^4.
\end{aligned} \tag{2.19}$$

Combining (2.18) and (2.19), we deduce that

$$\frac{c_0^2(q^2)}{c_0^2(q)} = (q^2; q^4)_\infty^2. \tag{2.20}$$

Iterating (2.20), we find that

$$\frac{c_0^2(q^{2^N})}{c_0^2(q)} = \prod_{\ell=1}^{N} \prod_{k=1}^{\infty} (1 - q^{2^\ell(2k-1)})^2.$$

Letting $N \to \infty$ and using the fact that $c_0(0) = 1$, we conclude that

$$\frac{1}{c_0^2(q)} = \prod_{k=1}^{\infty} (1 - q^{2k})^2,$$

where we have used the fact that any even integer is of the form $2^\ell(2k - 1)$. This completes the proof of (1.11).

Several existing proofs of (1.11), like that of the Borweins' proof, involve finding recurrence relations satisfied by the coefficients arising from the series representations in z of $(-zq; q^2)_\infty(-z^{-1}q; q^2)_\infty$. This method gives rise to (2.15) and one has to determine the constant term $c_0(q)$ to complete the proof of (1.11). The determination of $c_0(q)$ turns out to be the most difficult part of such proofs. See [48, p. 282] for another example of such proofs.

2.4 A Fermat-type equation and Jacobi's theta functions

It is well known [48, Section 13.2] that there are infinitely many integral solutions to the equation

$$x^2 + y^2 = z^2. \tag{2.21}$$

The solutions to the equation are called the Pythagorean triplets and are given by

$$x = k(a^2 - b^2),$$
$$y = 2kab,$$

and

$$z = k(a^2 + b^2),$$

with $k, a, b \in \mathbf{Z}$. However, when we replace the exponents of (2.21) by 4, the equation

$$x^4 + y^4 = z^4 \tag{2.22}$$

has no non-trivial (meaning $xyz \neq 0$) integral solutions [48, Section 13.3]. The following identity of Jacobi shows, however, that (2.22) holds when x, y and z are theta functions.

Theorem 2.4. *Let* $\Theta_2(q)$, $\Theta_3(q)$ *and* $\Theta_4(q)$ *be defined as in Definition 2.1. Then*

$$\Theta_3^4(q) = \Theta_2^4(q) + \Theta_4^4(q). \tag{2.23}$$

Proof. We first note that

$$\sum_{\substack{m,n\in\mathbf{Z} \\ \ell=m+ni \\ m\equiv n \,(\mathrm{mod}\,2)}} q^{N(\ell)} = \sum_{\substack{m,n\in\mathbf{Z} \\ \ell=m+ni \\ m\equiv n\equiv 0 \,(\mathrm{mod}\,2)}} q^{N(\ell)} + \sum_{\substack{m,n\in\mathbf{Z} \\ \ell=m+ni \\ m\equiv n\equiv 1 \,(\mathrm{mod}\,2)}} q^{N(\ell)} \tag{2.24}$$

and observe that

$$\sum_{\substack{m,n\in\mathbf{Z} \\ \ell=m+ni \\ m\equiv n\equiv 1 \,(\mathrm{mod}\,2)}} q^{N(\ell)} = \Theta_2^2(q^4)$$

and

$$\sum_{\substack{m,n\in\mathbf{Z} \\ \ell=m+ni \\ m\equiv n\equiv 0 \,(\mathrm{mod}\,2)}} q^{N(\ell)} = \Theta_3^2(q^4).$$

By (2.4), we find that

$$\sum_{\substack{m,n\in\mathbf{Z} \\ \ell=m+ni \\ m\equiv n \ (\text{mod } 2)}} q^{N(\ell)} = \sum_{r,s\in\mathbf{Z}} q^{N((1+i)r+(1-i)s)}$$

$$= \sum_{r,s\in\mathbf{Z}} q^{2(r^2+s^2)}$$

$$= \Theta_3^2(q^2). \tag{2.25}$$

Combining (2.24)–(2.25), we conclude that

$$\Theta_3^2(q^2) = \Theta_3^2(q^4) + \Theta_2^2(q^4). \tag{2.26}$$

Now, using (2.26), we find that

$$\Theta_3^2(q^4) - \Theta_2^2(q^4) = \Theta_3^2(q^4) - (\Theta_3^2(q^2) - \Theta_3^2(q^4))$$

$$= 2\Theta_3^2(q^4) - \Theta_3^2(q^2)$$

$$= \Theta_4^2(q^2), \tag{2.27}$$

where we have used (2.10) in the last equality. Multiplying (2.26) and (2.27) and using (2.1), we deduce that

$$\Theta_3^4(q^4) - \Theta_2^4(q^4) = \Theta_3^2(q^2)\Theta_4^2(q^2) = \Theta_4^4(q^4).$$

Replacing q^4 by q, we complete the proof of (2.23). □

If we study the proofs of all the identities up to this point, we would realize that the only identity that is hardest to prove is (2.25). This identity is also crucial in the proof of (2.1).

Identity (2.23) can be interpreted as solutions to (2.21), namely,

$$\left[\sum_{\ell=-\infty}^{\infty}\sum_{m=-\infty}^{\infty} q^{\ell^2+m^2}\right]^2 = \left[\sum_{\ell=-\infty}^{\infty}\sum_{m=-\infty}^{\infty} (-1)^{\ell+m} q^{\ell^2+m^2}\right]^2$$

$$+ \left[\sum_{\ell=-\infty}^{\infty}\sum_{m=-\infty}^{\infty} q^{(\ell+\frac{1}{2})^2+(m+\frac{1}{2})^2}\right]^2. \tag{2.28}$$

In 1991, Borwein and Borwein [14, (2.3)] discovered an amazing cubic analogue of (2.28). Their identity is given by

$$\left[\sum_{\ell=-\infty}^{\infty}\sum_{m=-\infty}^{\infty} q^{\ell^2+\ell m+m^2}\right]^3 = \left[\sum_{\ell=-\infty}^{\infty}\sum_{m=-\infty}^{\infty} e^{2\pi i(\ell-m)/3} q^{\ell^2+\ell m+m^2}\right]^3$$

$$+\left[\sum_{\ell=-\infty}^{\infty}\sum_{m=-\infty}^{\infty}q^{(\ell+\frac{1}{3})^2+(\ell+\frac{1}{3})(m+\frac{1}{3})+(m+\frac{1}{3})^2}\right]^3. \tag{2.29}$$

The proof of (2.29) can be found in [14].

2.5 A quotient of Jacobi's theta functions

In this section, we define an important function $\lambda(q)$ and derive a relation between $\lambda(q)$ and $\lambda(q^2)$ using Jacobi's identities satisfied by $\Theta_j(q)$, $j = 2, 3$ and 4.

Definition 2.5. Let $q \in \mathbf{C}$ with $|q| < 1$. The function $\lambda(q)$ is defined by

$$\lambda(q) = \frac{\Theta_2^4(q)}{\Theta_3^4(q)}. \tag{2.30}$$

In our proof of (2.23), we encountered the identities (2.26) and (2.27). Dividing (2.27) by (2.26) and replace q^2 by q, we deduce that

$$\frac{\Theta_3^2(q^2) - \Theta_2^2(q^2)}{\Theta_3^2(q^2) + \Theta_2^2(q^2)} = \frac{\Theta_4^2(q)}{\Theta_3^2(q)}.$$

This implies that

$$\frac{\left(1 - \frac{\Theta_2^2(q^2)}{\Theta_3^2(q^2)}\right)^2}{\left(1 + \frac{\Theta_2^2(q^2)}{\Theta_3^2(q^2)}\right)^2} = \frac{\Theta_4^4(q)}{\Theta_3^4(q)}$$

$$= \frac{\Theta_3^4(q) - \Theta_2^4(q)}{\Theta_3^4(q)}$$

$$= 1 - \frac{\Theta_2^4(q)}{\Theta_3^4(q)}, \tag{2.31}$$

where we have used (2.23) in the second last equality. Rearranging (2.31), we deduce the following formula.

Theorem 2.6. *Let $|q| < 1$ and $\lambda(q)$ be defined as in (2.30). Then*

$$\lambda(q^2) = \left(\frac{1 - \sqrt{1 - \lambda(q)}}{1 + \sqrt{1 - \lambda(q)}}\right)^2. \tag{2.32}$$

The notation $\sqrt{z}$ used in (2.32) denotes the principal branch of the multi-valued function $z^{1/2}$.

If we set $X = \lambda(q)$ and $Y = \lambda(q^2)$, we may rewrite (2.32) as

$$X^2 + X^2Y^2 - 2X^2Y - 16Y + 16XY = 0. \tag{2.33}$$

In general, it is known, from considering $\lambda(q)$ as a modular function, that when p is an odd prime, there is an identity of the form

$$P(\lambda(q), \lambda(q^p)) = 0,$$

where $P(x, y)$ is a homogeneous polynomial of 2 variables x and y with coefficients in $\mathbf{Z}$ and degree $p + 1$. An identity relating $\lambda(q)$ and $\lambda(q^p)$ is called a *modular equation of degree p*. Identity (2.33), which we derived using identities satisfied by Jacobi's theta functions, is a modular equation of degree 2. The construction of modular equations of degree p is tedious if we rely only on the properties of Jacobi's theta functions. Ramanujan constructed several modular equations for primes $3 \le p \le 23$. These identities and their proofs using theta functions can be found in Berndt's book [10, Chapter 19].

Exercises for Chapter 2

1. Derive the product representations of $\Theta_j(q), j = 2, 3$ and 4 using (1.11) and show that (2.1) holds.
2. Show that Jacobi's triple product identity implies (2.1).
3. Let $\lambda(q)$ be defined as in (2.5). Show that

$$\lambda(q^4) = \left(\frac{1 - \sqrt[4]{1 - \lambda(q)}}{1 + \sqrt[4]{1 - \lambda(q)}}\right)^4.$$

This is a modular equation of degree 4 associated with $\lambda(q)$.

4. Let $r_2(n)$ denote the number of ways of writing n as a sum of two squares. In other words,

$$\Theta_3^2(q) = \sum_{n=0}^{\infty} r_2(n)q^n.$$

   (a) Show that

$$r_2(n) = r_2(2n).$$

   Hint: Use the identity $2(a^2 + b^2) = (a + b)^2 + (a - b)^2$.
   (b) Deduce that

$$\Theta_3^2(q) + \Theta_4^2(q) = 2\Theta_3^2(q^2).$$

5. Let d be a positive integer such that $d \equiv 3 \pmod 4$. Show that

$$\Theta_3(q)\Theta_3(q^d) + \Theta_2(q)\Theta_2(q^d) = \sum_{m,n=-\infty}^{\infty} q^{m^2 + mn + (\frac{d+1}{4})n^2}.$$

6. (a) Let $f : \mathbf{Z} \times \mathbf{Z} \to \mathbf{C}$. Establish the formal identity

$$\sum_{m,n=-\infty}^{\infty} f(m, n) = \sum_{\ell,k=-\infty}^{\infty} f(\ell + k, \ell - k) + \sum_{\ell,k=-\infty}^{\infty} f(\ell + k, \ell - k - 1). \quad (2.34)$$

   (b) Use (2.34) to deduce that

$$\Theta_4(q)\Theta_4(q^3) + \Theta_2(q)\Theta_2(q^3) = \Theta_3(q)\Theta_3(q^3). \quad (2.35)$$

   Hint: You may need to use the identity

$$(h + j + 1)^2 + 3(h - j)^2 = \left(2h - j + \frac{1}{2}\right)^2 + 3\left(j + \frac{1}{2}\right)^2.$$

(c) Deduce that

$$\sqrt[4]{\lambda(q)\lambda(q^3)} + \sqrt[4]{(1-\lambda(q))(1-\lambda(q^3))} = 1.$$

(d) Find a homogeneous polynomial $P(x,y) \in \mathbf{Z}[x,y]$ of total degree 4 such that

$$P(\lambda(q), \lambda(q^3)) = 0.$$

7. The following problem is motivated by the proof of (1.11) given by Kongsiri-wong and Liu [56].

(a) Write

$$(q^2;q^2)_\infty(-zq;q^2)_\infty(-z^{-1}q,q^2)_\infty = \sum_{j=-\infty}^{\infty} a_j(q)z^j \qquad (2.36)$$

and show that

$$a_j(q) = a_{-j}(q)$$

and

$$a_j(q) = q^{2j-1}a_{j-1}(q), \quad j \ge 1.$$

Hence, deduce that

$$a_j(q) = q^{j^2} a_0(q).$$

(b) By replacing q with q^3 and letting $z = -q$ in (2.36), show that

$$(q^2;q^2)_\infty = a_0(q^3) \sum_{j=-\infty}^{\infty} (-1)^j q^{3j^2+j}.$$

(c) By replacing q with $q^{1/3}$ and letting $z = -q^{1/3}\omega$, where $\omega = e^{2\pi i/3}$ in (2.36), deduce that

$$(q^2;q^2)_\infty = a_0(q^{1/3}) \sum_{j=-\infty}^{\infty} (-1)^j q^{3j^2+j}.$$

(d) Deduce that

$$a_0(q) = a_0(q^9)$$

and complete the proof of (1.11).

8. (a) Recall from (1.14) that

$$(-z, q)_\infty = \sum_{j=0}^\infty \frac{q^{j(j-1)/2} z^j}{(q; q)_j}.$$

Use the above identity and the relation

$$(-z^3; q^3)_\infty = (-z; q)_\infty (-z\omega; q)_\infty (-z\omega^2; q)_\infty,$$

where $\omega = e^{2\pi i/3}$, to deduce that

$$\frac{q^{3k(k-1)/2}}{(q^3; q^3)_k} = \sum_{\substack{n_0 \geq 0, n_1 \geq 0, n_2 \geq 0 \\ n_0 + n_1 + n_2 = 3k}} \omega^{n_1 + 2n_2} \frac{q^{n_0(n_0-1)/2 + n_1(n_1-1)/2 + n_2(n_2-1)/2}}{(q; q)_{n_0} (q; q)_{n_1} (q; q)_{n_2}}.$$

(b) By letting $m_i = n_i - k$, show that

$$\frac{1}{(q^3; q^3)_k} = \sum_{\substack{m_0 \geq -k, m_1 \geq -k, m_2 \geq -k \\ m_0 + m_1 + m_2 = 0}} \omega^{m_1 + 2m_2} \frac{q^{(m_0^2 + m_1^2 + m_2^2)/2}}{(q; q)_{m_0 + k} (q; q)_{m_1 + k} (q; q)_{m_2 + k}}$$

and deduce that

$$\frac{(q; q)_\infty^3}{(q^3; q^3)_\infty} = \sum_{m_1=-\infty}^\infty \sum_{m_2=-\infty}^\infty \omega^{m_1 - m_2} q^{m_1^2 + m_1 m_2 + m_2^2}. \qquad (2.37)$$

The above method of deriving product representation of the series on the right-hand side of (2.37) is due to Borwein, Borwein and Garvan [15].

9. (a) Modify the method in Exercise 8 to show that

$$\frac{(q^3; q^3)_\infty^3}{(q; q)_\infty} = \sum_{m_1=-\infty}^\infty \sum_{m_2=-\infty}^\infty q^{3m_1^2 + 3m_2^2 + 3m_1 m_2 + m_1 + 2m_2}. \qquad (2.38)$$

(b) Show that

$$\sum_{m=-\infty}^\infty \sum_{\ell=-\infty}^\infty q^{(m+1/3)^2 + (m+1/3)(\ell+1/3) + (\ell+1/3)^2}$$

$$= 3q^{1/3} \sum_{m=-\infty}^\infty \sum_{\ell=-\infty}^\infty q^{3m^2 + 3\ell^2 + 3m\ell + m + 2\ell} \qquad (2.39)$$

and deduce from (2.38) that

$$\sum_{m=-\infty}^\infty \sum_{\ell=-\infty}^\infty q^{(m+1/3)^2 + (m+1/3)(\ell+1/3) + (\ell+1/3)^2} = 3q^{1/3} \frac{(q^3; q^3)_\infty^3}{(q; q)_\infty}.$$

Identity (2.39) can be found in [40, p. 181]. For other proofs of (2.39), see [74] or [36].

3 Two-variable extensions of Jacobi's theta functions and the partition function

Jacobi's theta functions introduced in Chapter 2 admit two-variable generalizations and we define them in this chapter. We then derive an important identity of Jacobi and use it to study Ramanujan's famous congruences satisfied by the partition function.

3.1 Two-variable extensions of Jacobi's theta functions

The product representations of $\Theta_j(q)$, $j = 2, 3$ and 4, introduced in Chapter 2 are consequences of the two-variable identity (1.11). It is therefore not surprising that these functions have two-variable extensions.

Definition 3.1. Let u and τ be complex numbers and $\operatorname{Im} \tau > 0$. Let $q = e^{\pi i \tau}$. *Jacobi's two-variable theta functions* are defined as follows:

$$
\begin{aligned}
\vartheta_1(u|\tau) &= -i \sum_{j=-\infty}^{\infty} (-1)^j q^{(j+\frac{1}{2})^2} e^{(2j+1)iu} \\
&= 2 \sum_{j=0}^{\infty} (-1)^j q^{(j+\frac{1}{2})^2} \sin((2j+1)u), \\
\vartheta_2(u|\tau) &= \sum_{j=-\infty}^{\infty} q^{(j+\frac{1}{2})^2} e^{(2j+1)iu} \\
&= 2 \sum_{j=0}^{\infty} q^{(j+\frac{1}{2})^2} \cos((2j+1)u), \\
\vartheta_3(u|\tau) &= \sum_{j=-\infty}^{\infty} q^{j^2} e^{2jiu} \\
&= 1 + 2 \sum_{j=1}^{\infty} q^{j^2} \cos(2ju),
\end{aligned}
\tag{3.1}
$$

and

$$
\begin{aligned}
\vartheta_4(u|\tau) &= \sum_{j=-\infty}^{\infty} (-1)^j q^{j^2} e^{2jiu} \\
&= 1 + 2 \sum_{j=1}^{\infty} (-1)^j q^{j^2} \cos(2ju).
\end{aligned}
$$

https://doi.org/10.1515/9783110541915-003

Observe that when $j = 2, 3, 4$ and $q = e^{\pi i \tau}$,

$$\vartheta_j(0|\tau) = \Theta_j(q).$$

Note that $\vartheta_1(0|\tau) = 0$ and this explains the absence of $\Theta_1(q)$ in the definition of Jacobi's theta functions of one variable given in Chapter 2.

One of the main applications of Jacobi's triple product identity (1.11) is to express Jacobi's two-variable functions in terms of infinite products. These identities are given as follows.

Theorem 3.2. *Let $u, \tau \in \mathbf{C}$ with $\operatorname{Im} \tau > 0$. Then*

$$\vartheta_1(u|\tau) = 2q^{1/4} \sin u \, (q^2; q^2)_\infty (e^{2iu} q^2; q^2)_\infty (e^{-2iu} q^2; q^2)_\infty, \qquad (3.2)$$

$$\vartheta_2(u|\tau) = 2q^{1/4} \cos u \, (q^2; q^2)_\infty (-e^{2iu} q^2; q^2)_\infty (-e^{-2iu} q^2; q^2)_\infty, \qquad (3.3)$$

$$\vartheta_3(u|\tau) = (q^2; q^2)_\infty (-e^{2iu} q; q^2)_\infty (-e^{-2iu} q; q^2)_\infty, \qquad (3.4)$$

$$\vartheta_4(u|\tau) = (q^2; q^2)_\infty (e^{2iu} q; q^2)_\infty (e^{-2iu} q; q^2)_\infty.$$

Proof. We will give the proof of (3.2) and leave the rest as exercises. To prove (3.2), we observe, after simplification, that

$$\vartheta_1(u|\tau) = -iq^{1/4} e^{iu} \sum_{j=-\infty}^{\infty} q^{j^2} (-qe^{2iu})^j.$$

By Jacobi's triple product identity (1.11) with $z = -qe^{2iu}$, we conclude that

$$\vartheta_1(u|\tau) = -iq^{1/4} e^{iu} (q^2; q^2)_\infty (e^{2iu} q^2; q^2)_\infty (e^{-2iu}; q^2)_\infty$$

$$= -iq^{1/4} e^{iu} (1 - e^{-2iu})(q^2; q^2)_\infty (e^{2iu} q^2; q^2)_\infty (e^{-2iu} q^2; q^2)_\infty$$

$$= 2q^{1/4} \sin u \, (q^2; q^2)_\infty (e^{2iu} q^2; q^2)_\infty (e^{-2iu} q^2; q^2)_\infty. \qquad \square$$

3.2 An important identity of Jacobi

Identity (3.2) shows that $\vartheta_1(u|\tau)$ has a simple zero at $u = 0$. We now determine $\vartheta_1'(0|\tau)$, where $f'(u|\tau)$ denotes the derivative of $f(u|\tau)$ with respect to u.

Differentiating both sides of (3.2) with respect to u, we find that

$$\vartheta_1'(u|\tau) = 2q^{1/4} (\cos u) T(u) + 2q^{1/4} (\sin u) T'(u),$$

where

$$T(u) = (q^2; q^2)_\infty (e^{2iu} q^2; q^2)_\infty (e^{-2iu} q^2; q^2)_\infty.$$

Since $\sin 0 = 0$ and

$$T(0) = (q^2; q^2)_\infty^3,$$

we conclude that

$$\vartheta_1'(0|\tau) = 2q^{1/4}(q^2; q^2)_\infty^3. \tag{3.5}$$

But the series expansion of $\vartheta_1(u|\tau)$ in (3.1) yields

$$\vartheta_1'(u|\tau) = 2\sum_{j=0}^{\infty}(-1)^j q^{(j+\frac{1}{2})^2}(2j+1)\cos(2j+1)u,$$

and this shows that

$$\vartheta_1'(0|\tau) = 2q^{1/4}\sum_{j=0}^{\infty}(2j+1)q^{j^2+j}. \tag{3.6}$$

Combining (3.5) and (3.6), and replacing q^2 by q, we deduce the following famous identity of Jacobi.

Corollary 3.3. *Let $q \in \mathbf{C}$ be such that $|q| < 1$. Then*

$$(q; q)_\infty^3 = \sum_{j=0}^{\infty}(-1)^j(2j+1)q^{j(j+1)/2}. \tag{3.7}$$

3.3 The partition function $p(n)$ and its generating function

Definition 3.4. A partition of a positive integer n is a representation of n as a sum of non-decreasing positive integers, called summands or parts of the partition. The partition function $p(n)$ is defined as the number of partitions of n.

The function $p(n)$ is not defined for $n = 0$ but we will set $p(0) = 1$.

Example 3.1. The integer 4 has the following partitions:

$$4 = 1 + 3 = 2 + 2 = 1 + 1 + 2 = 1 + 1 + 1 + 1.$$

The number of partitions of 4 is 5 and we write $p(4) = 5$.

One of the most important identities associated with $p(n)$ is the following.

Theorem 3.5. *For $|q| < 1$, we have*

$$\frac{1}{(q; q)_\infty} = \sum_{n=0}^{\infty}p(n)q^n. \tag{3.8}$$

Proof. We will follow the proof outlined in [48, Section 19.3]. Suppose that q is real and $0 < q < 1$ so that the product

$$F(q) = \frac{1}{(q;q)_\infty} = \prod_{k=1}^\infty \frac{1}{1 - q^k}$$

is convergent. Next, for $0 < q < 1$, the coefficient of q^n in the expression

$$F_m(q) = \prod_{k=1}^m \frac{1}{1 - q^k} = \prod_{k=1}^m \sum_{j=0}^\infty q^{kj}$$

is $p_m(n)$, the number of partitions of $n \geq 1$ into parts not exceeding m. Setting $p_m(0) = 1$, we find that

$$F_m(q) = \sum_{n=0}^\infty p_m(n)q^n. \tag{3.9}$$

Since $p(n)$ counts the number of partitions of n without restrictions on the size of its parts, we conclude that

$$p_m(n) \leq p(n) \tag{3.10}$$

for all positive integers m with equality when $n \leq m$.

Next, we write

$$F_m(q) = \sum_{n=0}^m p(n)q^n + \sum_{n=m+1}^\infty p_m(n)q^n.$$

Note that $F_m(q) \leq F(q)$ and

$$\lim_{m\to\infty} F_m(q) = F(q).$$

Thus,

$$\sum_{n=0}^m p(n)q^n = \sum_{n=0}^m p_m(n)q^n < F_m(q) < F(q).$$

This implies that, for $0 < q < 1$, the series

$$\sum_{n=0}^\infty p(n)q^n$$

is convergent. By (3.10), we deduce that

$$\sum_{n=0}^m p_m(n)q^n \leq \sum_{n=0}^\infty p(n)q^n,$$

and this implies that, for $0 < q < 1$,

$$\sum_{n=0}^{\infty} p_m(n)q^n$$

converges uniformly for all values of m. Since

$$\lim_{m\to\infty} p_m(n) = p(n),$$

we deduce from Lemma 1.6 and (3.9) that

$$\sum_{n=0}^{\infty} p(n)q^n = \lim_{m\to\infty} \sum_{n=0}^{\infty} p_m(n)q^n = \lim_{m\to\infty} F_m(q) = F(q). \qquad (3.11)$$

The proof of the theorem is complete using analytic continuation by observing that both sides of (3.11) are analytic functions on $|q| < 1$. $\qquad\square$

3.4 The values of $p(n)$

Observe from Theorem 3.5 that

$$1 = (q;q)_\infty \sum_{n=0}^{\infty} p(n)q^n.$$

Using (1.12), we find that

$$1 = \left(1 + \sum_{v=1}^{\infty}(-1)^v(q^{v(3v+1)/2} + q^{v(3v-1)/2})\right) \sum_{j=0}^{\infty} p(j)q^j$$

$$= \left(1 + \sum_{\mu=0}^{\infty}(-1)^{\mu+1}(q^{(\mu+1)(3\mu+2)/2}) + q^{(\mu+1)(3\mu+4)/2}\right) \sum_{j=0}^{\infty} p(j)q^j.$$

Hence, we deduce that, for $N \geq 1$,

$$p(N) = \sum_{\substack{j,\mu\geq 0 \\ j+\frac{(\mu+1)(3\mu+2)}{2}=N}} (-1)^\mu p(j) + \sum_{\substack{j,\mu\geq 0 \\ j+\frac{(\mu+1)(3\mu+4)}{2}=N}} (-1)^\mu p(j). \qquad (3.12)$$

Major MacMahon, using (3.12), tabulated $p(n)$ up to $n = 200$.

Ramanujan was the first mathematician to notice that there might be a formula for determining $p(n)$ directly without using the recurrence (3.12). Around 1918, Ramanujan and Hardy [47] succeeded in expressing $p(n)$ as the integer part of a finite sum which diverges as the upper index of the summation tends to ∞.

In 1937, Rademacher discovered an exact formula of $p(n)$ in terms of a convergent series. Selberg had also independently discovered the same formula around the same time. For more details about Selberg's independent discovery of the exact formula of $p(n)$, the readers are encouraged to read Selberg's article [66, p. 701–706].

We end this section by giving two formulas relating $p(n)$ and $\sigma(n)$, where for $n \geq 1$,

$$\sigma(n) = \sum_{\ell \mid n} \ell.$$

We will set $\sigma(0) = 0$. By logarithmically differentiating (3.8), we find that

$$\sum_{n=0}^{\infty} np(n)q^n = \sum_{v=1}^{\infty} \frac{vq^v}{1-q^v} \sum_{\mu=0}^{\infty} p(\mu)q^\mu.$$

Using the fact that

$$\sum_{v=1}^{\infty} \frac{vq^v}{1-q^v} = \sum_{v=1}^{\infty} \sum_{m=1}^{\infty} vq^{mv}$$

$$= \sum_{s=1}^{\infty} \sigma(s)q^s,$$

we deduce that

$$np(n) = \sum_{s+\mu=n} \sigma(s)p(\mu). \tag{3.13}$$

This is the first relation between $p(n)$ and $\sigma(n)$.

Next, instead of using (3.8), we differentiate

$$G(q) = \prod_{n=1}^{\infty}(1-q^n)$$

logarithmically to deduce that

$$q\frac{G'(q)}{G(q)} = -\sum_{j=1}^{\infty} \sigma(j)q^j. \tag{3.14}$$

Using (1.12), we write

$$qG'(q) = \sum_{\ell=1}^{\infty}(-1)^\ell\left(\frac{\ell(3\ell+1)}{2}q^{\ell(3\ell+1)/2} + \frac{\ell(3\ell-1)}{2}q^{\ell(3\ell-1)/2}\right).$$

Furthermore, by using (3.8), we conclude from (3.14) that

$$\sigma(n) = \sum_{\substack{j,\ell \geq 0 \\ j + \frac{(\ell+1)(3\ell+4)}{2} = n}} (-1)^\ell \frac{(\ell + 1)(3\ell + 4)}{2} p(j)$$

$$+ \sum_{\substack{j,\ell \geq 0 \\ j + \frac{(\ell+1)(3\ell+2)}{2} = n}} (-1)^\ell \frac{(\ell + 1)(3\ell + 2)}{2} p(j). \qquad (3.15)$$

This is the second relation between $p(n)$ and $\sigma(n)$ and it was discovered by Osler, Hassen and Chandrupatia [61, p. 287]. It is now immediate that we can tabulate both $p(n)$ and $\sigma(n)$ using (3.13) and (3.15).

Example 3.2. We begin with $p(1) = 1$ and $\sigma(1) = 1$. To compute $p(2)$, we need the value of $\sigma(2)$. Of course, we know that $\sigma(2) = 3$. But we can also obtain the value of $\sigma(2)$ from (3.15) since

$$\sigma(2) = \frac{1 \cdot 4}{2} p(0) + \frac{1 \cdot 2}{2} p(1) = 3.$$

Using this value, we compute $p(2)$ using (3.13), namely,

$$2p(2) = \sigma(1)p(1) + \sigma(2)p(0) = 4,$$

which gives $p(2) = 2$. We then determine $\sigma(3)$ using (3.15) and repeat the process to tabulate the values of $p(n)$ and $\sigma(n)$.

Since n is a prime if and only if n has exactly two divisors, we immediately know that n is a prime if $\sigma(n) = 1 + n$. We illustrate this using $n = 19$. To compute $\sigma(19)$, we need the values $p(17) = 297, p(12) = 77, p(4) = 5, p(18) = 385, p(14) = 135$ and $p(7) = 15$. By (3.15), we conclude that

$$\sigma(19) = 2 \cdot 297 - 7 \cdot 77 + 15 \cdot 5 + 385 - 5 \cdot 135 + 12 \cdot 15 = 20$$

and hence 19 is a prime. Identity (3.15) shows that we can determine if an integer n were prime by computing the values of $p(m)$ for certain m between 1 and n. To determine if n were a prime for large n, the values of $p(m)$ for $1 \leq m < n$ which we need may not be efficiently determined using (3.13). We could instead evaluate these values using (3.12) or the Ramanujan–Hardy–Rademacher formula for $p(m)$ mentioned after the derivation of (3.12).

3.5 Ramanujan's congruences for $p(n)$

According to Hardy [48, p. 287], Ramanujan was led to three striking congruences satisfied by $p(n)$ when examining MacMahon's table of $p(n)$.

Ramanujan's first congruence is stated in the following theorem.

Theorem 3.6. *Let n be a positive integer. Then*

$$p(5n - 1) \equiv 0 \pmod{5}. \tag{3.16}$$

Proof. Write

$$q \prod_{k=1}^{\infty} (1 - q^k)^4 = q \prod_{k=1}^{\infty} (1 - q^k) \prod_{k=1}^{\infty} (1 - q^k)^3$$

$$= \sum_{r=-\infty}^{\infty} \sum_{s=0}^{\infty} (-1)^{r+s} (2s + 1) q^{1+r(3r+1)/2+s(s+1)/2},$$

where we have used (1.12) and (3.7). Observe that when

$$1 + r(3r + 1)/2 + s(s + 1)/2$$

is a multiple of 5, then

$$2 + 3r^2 + r + s^2 + s \equiv 0 \pmod{5}.$$

Multiplying both sides by 4, we conclude that

$$3 + 2r^2 + 4r + 4s^2 + 4s \equiv 0 \pmod{5},$$

which implies that

$$2(r + 1)^2 + (2s + 1)^2 \equiv 0 \pmod{5}. \tag{3.17}$$

Now, we find that

$$2M^2 + N^2 \equiv 0 \pmod{5}$$

has a solution if

$$N^2 \equiv -2M^2 \pmod{5}. \tag{3.18}$$

If $M \not\equiv 0 \pmod{5}$, then (3.18) implies that NM^{-1} is a solution of

$$x^2 \equiv -2 \pmod{5}.$$

This is false since -2 is not a quadratic residue modulo 5. Hence, if (3.18) holds, we must have

$$M \equiv 0 \pmod{5}$$

and

$$N \equiv 0 \pmod 5.$$

Returning to (3.17), we observe that if (3.17) holds, then

$$2s + 1 \equiv 0 \pmod 5. \tag{3.19}$$

If we write

$$q\prod_{n=1}^{\infty}(1 - q^n)^4 = \sum_{j=0}^{\infty} a_j q^j,$$

then the coefficient of q^{5j} in the above expansion is given by

$$a_{5j} = \sum_{\substack{r,s=-\infty \\ 1+r(3r+1)/2+s(s+1)/2=5j}}^{\infty} (-1)^{r+s}(2s + 1) \equiv 0 \pmod 5, \tag{3.20}$$

where our last congruence follows from (3.19).

Next, we recall that if $f(q)$ and $g(q)$ are two power series with integral coefficients, then for any prime p, we write

$$f(q) \equiv g(q) \pmod p$$

if

$$f(q) - g(q) = ph(q)$$

for some power series $h(q)$ with integral coefficients. Observe that

$$1 - q^{5n} \equiv (1 - q^n)^5 \pmod 5$$

and this implies that

$$q\prod_{n=1}^{\infty}\frac{(1 - q^{5n})}{(1 - q^n)} \equiv q\prod_{n=1}^{\infty}(1 - q^n)^4 \pmod 5.$$

Thus, by (3.8), we deduce that

$$\sum_{k=0}^{\infty} p(k)q^{k+1} \equiv \sum_{j=0}^{\infty} a_j q^j \prod_{\ell=1}^{\infty}\frac{1}{1 - q^{5\ell}}$$

$$\equiv \sum_{j=0}^{\infty} a_j q^j \sum_{\ell=0}^{\infty} p(\ell)q^{5\ell} \pmod 5. \tag{3.21}$$

Comparing coefficients of q^{5m} on both sides of (3.21), we conclude that

$$p(5m - 1) \equiv \sum_{5j+5\ell=5m} a_{5j}p(\ell) \equiv 0 \pmod{5},$$

where the last congruence follows from (3.20). □

Besides (3.16), Ramanujan also discovered the congruences

$$p(7n - 2) \equiv 0 \pmod{7} \tag{3.22}$$

and

$$p(11n - 5) \equiv 0 \pmod{11}. \tag{3.23}$$

The proof of (3.22) is similar to the proof of (3.16) and we leave the details as an exercise.

Remark 3.1. In order to give a proof similar to Ramanujan's method for (3.23), one would need to consider the series expansion of $\prod_{k=1}^{\infty}(1-q^k)^{10}$. Ramanujan did not have a series representation of the above product. This was provided by Winquist [73]. He showed that

$$\prod_{k=1}^{\infty}(1 - q^k)^{10} = \sum_{v=-\infty}^{\infty} \sum_{\mu=-\infty}^{\infty} (-1)^{v+\mu}(2v + 1)(6\mu + 1)$$

$$\times \left(\frac{(3v + 1)(3v + 2)}{4} - \frac{3\mu(3\mu + 1)}{4} \right) q^{3v(v+1)/2+\mu(3\mu+1)/2} \tag{3.24}$$

and provided the first published proof of (3.23) in the spirit of Ramanujan's proof of (3.16) and (3.22). Winquist's identity (3.24) can be interpreted as a special case of Macdonald's identity for twisted affine Kac–Moody algebras. For more details, see Carter's book [21, Theorem 20.4 and p. 491]. For a discussion of (3.24) using elliptic functions, see [30, p. 233].

Remark 3.2. Although Ramanujan did not give a proof of (3.23) using product representation of $\prod_{k=1}^{\infty}(1 - q^k)^{10}$, he gave a proof of (3.23) using another method. For more details, see [64, p. 147].

Remark 3.3. In 2003, Boylan and Alghren [4] showed that, for any integer $n \geq 0$,

$$p(\ell n - \delta_\ell) \equiv 0 \pmod{\ell}$$

for some integer δ_ℓ if and only if

$$\ell = 5, 7 \text{ and } 11.$$

Remark 3.4. In 1996, Eichhorn and Ono [43] showed that if 5 divides $p(4)$, $p(14)$ and $p(19)$, then (3.16) holds. Similarly, (3.22) follows from the divisibility of first 5 values of $p(7n-2)$ by 7 and (3.23) follows from the divisibility of the first 11 values of $p(11n-5)$ by 11. In [31], Chan and Lewis showed that the number of congruences to be verified can be reduced to 1, 2 and 5 for (3.16), (3.22) and (3.23) respectively. Recently, Chan, Wang and Yang [37] showed that Ramanujan's congruences for 5, 7 and 11 follow from the facts that $p(4) = 5$, $p(5) = 7$ and $p(6) = 11$, respectively. This means that in order to establish Ramanujan's congruences (3.16), (3.22) and (3.23), only the first value of $p(n)$ in the respective arithmetic progression is needed.

Exercises for Chapter 3

1. Let n be a positive integer. Show that

    $$p(7n - 2) \equiv 0 \;(\mathrm{mod}\; 7)$$

    by first considering

    $$q^2 \prod_{n=1}^{\infty}(1 - q^n)^6$$

    and its series expansion using

    $$\prod_{n=1}^{\infty}(1 - q^n)^3 = \sum_{n=0}^{\infty}(-1)^n(2n + 1)q^{n(n+1)/2}.$$

2. Use (3.24) to deduce that

    $$p(11n - 5) \equiv 0 \;(\mathrm{mod}\; 11)$$

    for $n \geq 1$.

3. Show that

    $$\vartheta_1'(0|\tau) = \vartheta_2(0|\tau)\vartheta_3(0|\tau)\vartheta_4(0|\tau),$$

    by using the product representations of $\vartheta_1'(0|\tau)$, $\vartheta_2(0|\tau)$, $\vartheta_3(0|\tau)$ and $\vartheta_4(0|\tau)$.

4. Let S be a subset of the set of positive integers. Let $p(S, n)$ denote the number of partitions of n in which each summand is an element of S. For integer $d > 2$, let S_d denote the set of all positive integers congruent to 1 or to -1 modulo $d+3$. Show that

    $$\sum_{j=0}^{\infty} p(S_1, j)q^j = \prod_{k=1}^{\infty}\frac{1}{1 - q^{2k-1}},$$

    $$\sum_{j=0}^{\infty} p(S_2, j)q^j = \prod_{k=1}^{\infty}\frac{1}{(1 - q^{5k-1})(1 - q^{5k-4})},$$

    and

    $$\sum_{j=0}^{\infty} p(S_3, j)q^j = \prod_{k=1}^{\infty}\frac{1}{(1 - q^{6k-1})(1 - q^{6k-5})}.$$

5. If $Q_3(n)$ denotes the number of partitions of n into distinct parts, no part being a multiple of 3, show that $Q_3(n) = p(S_3, n)$ for all $n \in \mathbf{Z}^+$.

4 Ramanujan's differential equations

In this chapter, we encounter the Lambert series $L_{2j}(q), j \geq 1$ and derive an important system of differential equations satisfied by $L_2(q)$, $L_4(q)$ and $L_6(q)$ using Jacobi's triple product identity and the quintuple product identity.

4.1 The Bernoulli numbers and cot u

We begin this section by studying the power series expansion of $\cot u$. To describe the coefficients of the series representation of $\cot u$, we need the Bernoulli numbers.

Definition 4.1. For $|x| < 2\pi$, let

$$\frac{x}{e^x - 1} = \sum_{j=0}^{\infty} B_j \frac{x^j}{j!}. \tag{4.1}$$

The numbers B_j are called the *Bernoulli numbers*.

The first few values of B_n are displayed as follows:

Table 4.1: Examples of Bernoulli numbers.

n	0	1	2	3	4	5	6	7	8	9	10	11	12
B_n	1	$-\frac{1}{2}$	$\frac{1}{6}$	0	$-\frac{1}{30}$	0	$\frac{1}{42}$	0	$-\frac{1}{30}$	0	$\frac{5}{66}$	0	$-\frac{691}{2730}$

Now, write

$$\cot u = \frac{\cos u}{\sin u}$$

$$= i\left(\frac{e^{iu} + e^{-iu}}{e^{iu} - e^{-iu}}\right)$$

$$= i\left(\frac{e^{2iu} + 1}{e^{2iu} - 1}\right)$$

$$= i\left(1 + \frac{2}{e^{2iu} - 1}\right).$$

Using the expansion (4.1), we conclude that

$$\frac{1}{e^{2iu} - 1} = \sum_{j=0}^{\infty} \frac{B_j}{j!}(2iu)^{j-1}.$$

https://doi.org/10.1515/9783110541915-004

Therefore,

$$\cot u = i + \sum_{j=0}^{\infty} \frac{B_j}{j!}(2i)^j u^{j-1}$$

$$= \frac{1}{u} + \frac{1}{u}\sum_{j=2}^{\infty} \frac{B_j}{j!}(2i)^j u^j, \tag{4.2}$$

where we have used the fact that $B_1 = -1/2$ to remove the constant term. Since $\cot u$ is an odd function, we conclude that when $j \geq 1$,

$$B_{2j+1} = 0. \tag{4.3}$$

Using (4.3), we may rewrite (4.2) as

$$\cot u = \frac{1}{u} + \sum_{j=2}^{\infty} (-1)^j \frac{B_{2j}}{(2j)!} 2^{2j} u^{2j-1}. \tag{4.4}$$

It can be shown that, for every positive integer j, $B_{2j} \neq 0$. This follows from the formulae (see Exercise 1 at the end of this chapter). We have

$$\sum_{k=1}^{\infty} \frac{1}{k^{2j}} = \frac{(-1)^{j+1}B_{2j}(2\pi)^{2j}}{2 \cdot (2j)!}. \tag{4.5}$$

4.2 The logarithmic derivative of $\vartheta_1(u|\tau)$

In (3.2), we encountered the formula

$$\vartheta_1(u|\tau) = 2q^{1/4}\sin u \prod_{k=1}^{\infty}(1 - q^{2k})(1 - q^{2k}e^{2iu})(1 - q^{2k}e^{-2iu}). \tag{4.6}$$

Logarithmically differentiating (4.6) with respect to u, we deduce that

$$\frac{\vartheta_1'(u|\tau)}{\vartheta_1(u|\tau)} = \cot u + 2i\left(\sum_{k=1}^{\infty} \frac{q^{2k}e^{-2iu}}{1 - q^{2k}e^{-2iu}} - \sum_{k=1}^{\infty} \frac{q^{2k}e^{2iu}}{1 - q^{2k}e^{2iu}}\right)$$

$$= \cot u + 2i\sum_{\ell=1}^{\infty}\sum_{k=1}^{\infty} q^{2k\ell}(e^{-2iu\ell} - e^{2iu\ell})$$

$$= \cot u + 4\sum_{\ell=1}^{\infty} \frac{q^{2\ell}}{1 - q^{2\ell}} \sin 2\ell u. \tag{4.7}$$

Using (4.4) and the expansion for $\sin u$, we deduce that

$$\frac{\vartheta_1'(u|\tau)}{\vartheta_1(u|\tau)} = \frac{1}{u} + \sum_{j=1}^{\infty} \frac{(-1)^j 2^{2j} B_{2j}}{(2j)!} u^{2j-1} + 4\sum_{j=1}^{\infty}\left(\sum_{\ell=1}^{\infty} \frac{q^{2\ell}}{1 - q^{2\ell}} \ell^{2j-1}\right)\frac{(-1)^{j-1} 2^{2j-1}}{(2j-1)!} u^{2j-1}.$$

Hence, we deduce the following theorem.

Theorem 4.2. *Let $j \in \mathbf{Z}^+$, $u, \tau \in \mathbf{C}$ with $\operatorname{Im}\tau > 0$ and $q = e^{\pi i \tau}$. If*

$$L_{2j}(x) = 1 - \frac{4j}{B_{2j}} \sum_{\ell=1}^{\infty} \frac{\ell^{2j-1} x^{\ell}}{1 - x^{\ell}}, \tag{4.8}$$

then

$$\frac{\vartheta_1'(u|\tau)}{\vartheta_1(u|\tau)} = \frac{1}{u} + \sum_{j=1}^{\infty} (-1)^j \frac{2^{2j}}{(2j)!} B_{2j} L_{2j}(q^2) u^{2j-1}. \tag{4.9}$$

We note that the division by B_{2j} in the definition of $L_{2j}(x)$ is justified since $B_{2j} \neq 0$ by (4.5).

For $j = 1, 2, 3$ and 4, we have

$$L_2(x) = 1 - 24 \sum_{\ell=1}^{\infty} \frac{\ell x^{\ell}}{1 - x^{\ell}},$$

$$L_4(x) = 1 + 240 \sum_{\ell=1}^{\infty} \frac{\ell^3 x^{\ell}}{1 - x^{\ell}},$$

$$L_6(x) = 1 - 504 \sum_{\ell=1}^{\infty} \frac{\ell^5 x^{\ell}}{1 - x^{\ell}},$$

and

$$L_8(x) = 1 + 480 \sum_{\ell=1}^{\infty} \frac{\ell^7 x^{\ell}}{1 - x^{\ell}},$$

where we have used the values of B_{2k} given in Table 4.1.

Our aim is to prove three important differential equations satisfied by $L_2(x)$, $L_4(x)$ and $L_6(x)$. Before we discuss these differential equations, we need another identity that is equivalent to the quintuple product identity.

Theorem 4.3. *Let $u, q \in \mathbf{C}$ with $|q| < 1$. Then*

$$q^{1/12} \left(\prod_{k=1}^{\infty} (1 - q^{2k}) \right) \frac{\vartheta_1(2u|\tau)}{\vartheta_1(u|\tau)} = 2 \sum_{j=-\infty}^{\infty} (-1)^j q^{(6j+1)^2/12} \cos(6j+1)u. \tag{4.10}$$

Proof. Since

$$\vartheta_1(u|\tau) = 2q^{1/4} \sin u \prod_{k=1}^{\infty} (1 - q^{2k})(1 - q^{2k} e^{2iu})(1 - q^{2k} e^{-2iu}),$$

we find that

$$\frac{\vartheta_1(2u|\tau)}{\vartheta_1(u|\tau)} = \frac{\sin 2u}{\sin u} \prod_{k=1}^{\infty} \frac{(1 - q^{2k}e^{4iu})(1 - q^{2k}e^{-4iu})}{(1 - q^{2k}e^{2iu})(1 - q^{2k}e^{-2iu})}$$

$$= 2\cos u \prod_{k=1}^{\infty} \frac{(1 - q^k e^{2iu})(1 + q^k e^{2iu})(1 + q^k e^{-2iu})(1 - q^k e^{-2iu})}{(1 - q^{2k}e^{2iu})(1 - q^{2k}e^{-2iu})}$$

$$= 2\cos u \prod_{k=1}^{\infty} (1 - q^{2k-1}e^{2iu})(1 - q^{2k-1}e^{-2iu})(1 + q^k e^{2iu})(1 + q^k e^{-2iu})$$

$$= 2\cos u \prod_{k=1}^{\infty} \frac{(1 - q^{4k-2}e^{4iu})(1 - q^{4k-2}e^{-4iu})}{(1 + q^{2k-1}e^{2iu})(1 + q^{2k-1}e^{-2iu})}(1 + q^k e^{2iu})(1 + q^k e^{-2iu})$$

$$= 2\cos u \prod_{k=1}^{\infty} (1 - q^{4k-2}e^{4iu})(1 - q^{4k-2}e^{-4iu})(1 + q^{2k}e^{2iu})(1 + q^{2k}e^{-2iu}).$$

Using the quintuple product identity (1.22) with $t = -e^{2iu}$, we deduce that

$$2\cos u \prod_{k=1}^{\infty} (1 - q^{2k})(1 + q^{2k}e^{2iu})(1 + q^{2k}e^{-2iu})(1 - q^{4k-2}e^{4iu})(1 - q^{4k-2}e^{-4iu})$$

$$= e^{iu} \sum_{j=-\infty}^{\infty} q^{3j^2+j}(e^{6iuj} + e^{-6iuj-2iu})(-1)^j$$

$$= 2 \sum_{j=-\infty}^{\infty} (-1)^j q^{3j^2+j} \cos(6j+1)u.$$

This completes the proof of Theorem 4.3. □

4.3 Ramanujan's differential equations for $L_2(x)$, $L_4(x)$ and $L_6(x)$

In [63, (30)], Ramanujan derived a system of three differential equations involving $L_2(x)$, $L_4(x)$ and $L_6(x)$, as stated in the following theorem.

Theorem 4.4. *For $x \in \mathbf{C}$ with $|x| < 1$, we have*

$$x\frac{dL_2(x)}{dx} = \frac{L_2^2(x) - L_4(x)}{12}, \tag{4.11}$$

$$x\frac{dL_4(x)}{dx} = \frac{L_2(x)L_4(x) - L_6(x)}{3}, \tag{4.12}$$

and

$$x\frac{dL_6(x)}{dx} = \frac{L_2(x)L_6(x) - L_4^2(x)}{2}. \tag{4.13}$$

Ramanujan used (4.11)–(4.13) to derive perhaps the most elementary proof of the following identity.

Theorem 4.5. *Let $x \in \mathbf{C}$ with $|x| < 1$. If*

$$D(x) = x \prod_{k=1}^{\infty} (1 - x^k)^{24},$$ (4.14)

then

$$1728 D(x) = L_4^3(x) - L_6^2(x).$$ (4.15)

The first mathematician who discovered the differential equations (4.11)–(4.12) appears to be Halphen (see [46, p. 450] and [40, p. 43]). These differential equations are now known as Ramanujan's differential equations. They are also featured early in Swinnerton-Dyer's theory of modular forms modulo p (see [69] or [57, Theorem 5.3]).

Before we prove Theorem 4.4, we first show how Ramanujan deduced Theorem 4.5 from Theorem 4.4.

Proof of Theorem 4.5. First, note that by (4.12) and (4.13),

$$3L_4^2(x)x\frac{dL_4(x)}{dx} - 2L_6(x)x\frac{dL_6(x)}{dx}$$
$$= L_4^2(x)(L_2(x)L_4(x) - L_6(x)) - L_6(x)(L_2(x)L_6(x) - L_4^2(x))$$
$$= L_2(x)(L_4^3(x) - L_6^2(x)).$$

This implies that

$$x\frac{d\ln(L_4^3(x) - L_6^2(x))}{dx} = L_2(x) = x\frac{d\ln D(x)}{dx},$$

where we have used the identity

$$x\frac{d\ln D(x)}{dx} = L_2(x)$$

in the last equality. Hence,

$$L_4^3(x) - L_6^2(x) = cD(x)$$

for some constant c. By considering the coefficient of x in $D(x)$ and $L_4^3(x) - L_6^2(x)$, we find that $c = 1728$ and this completes the proof of the identity. $\square$

We now proceed with a proof of Theorem 4.4. We will follow the approach given in Chan's article [26]. For an alternative approach given by Ramanujan, see [63].

Proof of Theorem 4.4. Let

$$S_{2n+1} = 2 \sum_{k=0}^{\infty} (-1)^k (2k+1)^{2n+1} q^{(2k+1)^2/4}.$$

By expanding

$$\vartheta_1(u|\tau) = 2 \sum_{k=0}^{\infty} (-1)^k q^{(2k+1)^2/4} \sin(2k+1)u$$

in a power series in u, we find that

$$\vartheta_1(u|\tau) = \sum_{j=0}^{\infty} \frac{(-1)^j}{(2j+1)!} S_{2j+1} u^{2j+1}.$$

From (4.9), we deduce that

$$\sum_{j=0}^{\infty} \frac{(-1)^j}{(2j)!} S_{2j+1} u^{2j} = \left(\sum_{j=0}^{\infty} \frac{(-1)^j}{(2j+1)!} S_{2j+1} u^{2j+1} \right) \left(\frac{1}{u} + \sum_{j=1}^{\infty} (-1)^j \frac{2^{2j} B_{2j}}{(2j)!} L_{2j}(q^2) u^{2j-1} \right).$$

Comparing the coefficients of u^{2n+1} of both sides of the above identity, we find that

$$S_{2n+1} = \frac{1}{2n} \sum_{j=1}^{n} 2^{2j} \binom{2n+1}{2j} B_{2j} L_{2j}(q^2) S_{2(n-j)+1}.$$

We record the first four identities

$$S_3 = S_1 L_2(q^2), \tag{4.16}$$

$$S_5 = S_1 \frac{5L_2^2(q^2) - 2L_4(q^2)}{3}, \tag{4.17}$$

$$S_7 = S_1 \frac{35L_2^3(q^2) - 42L_2(q^2)L_4(q^2) + 16L_6(q^2)}{9}, \tag{4.18}$$

and

$$S_9 = S_1 \left(\frac{35}{3} L_2^4(q^2) - 28L_2^2(q^2)L_4(q^2) + \frac{64}{3} L_2(q^2)L_6(q^2) + \frac{28}{5} L_4^2(q^2) - \frac{48}{5} L_8(q^2) \right). \tag{4.19}$$

Now, applying $q\frac{d}{dq}$ to (4.16), we deduce that

$$S_5 = 4q\frac{dL_2(q^2)}{dq}S_1 + L_2(q^2)S_3,$$

where we have used the identity

$$4q\frac{dS_{2n+1}}{dq} = S_{2n+3},$$

which holds for any positive integer n. This gives

$$S_5 = S_1\left(4q\frac{dL_2(q^2)}{dq} + L_2^2(q^2)\right).$$

Comparing this with (4.17), we deduce that

$$q\frac{dL_2(q^2)}{dq} = \frac{L_2^2(q^2) - L_4(q^2)}{6},$$

which, upon using $x = q^2$, leads to (4.11).

Similarly by using the same idea on (4.17) and (4.18), (4.18) and (4.19), we deduce

$$q\frac{dL_4(q^2)}{dq} = \frac{2(L_2(q^2)L_4(q^2) - L_6(q^2))}{3} \tag{4.20}$$

and

$$-112L_4^2(q^2) - 320L_2(q^2)L_6(q^2) + 320q\frac{dL_6(q^2)}{dq} + 432L_8(q^2) = 0. \tag{4.21}$$

Note that substituting $x = q^2$ in (4.20) yields (4.12).

It remains to establish (4.13). To prove (4.13), we need another relation similar to (4.21).

In order to derive such a relation, we rewrite (4.10) as

$$q^{1/12}\prod_{k=1}^{\infty}(1 - q^{2k})\frac{\vartheta_1(2u|\tau)}{\vartheta_1(u|\tau)} = \sum_{j=0}^{\infty}(-1)^j\frac{T_{2j}}{(2j)!}u^{2j}, \tag{4.22}$$

where

$$T_{2j} = 2\sum_{\ell=-\infty}^{\infty}(-1)^{\ell}(6\ell + 1)^{2j}q^{(6\ell+1)^2/12}.$$

By logarithmically differentiating (4.22) with respect to u, we deduce that

$$\left(2\frac{\vartheta_1'}{\vartheta_1}(2u|\tau) - \frac{\vartheta_1'}{\vartheta_1}(u|\tau)\right)\left(\sum_{j=0}^{\infty}(-1)^j\frac{T_{2j}}{(2j)!}u^{2j}\right) = \sum_{j=0}^{\infty}(-1)^{j+1}\frac{T_{2j+2}}{(2j+1)!}u^{2j+1}.$$

Hence, we find that, for positive integer n,

$$T_{2n+2} = \frac{1}{(2n+2)}\sum_{j=0}^{n}\binom{2n+2}{2j+2}2^{2j+2}(2^{2j+2}-1)B_{2j+2}L_{2j+2}(q^2)T_{2(n-j)}.$$

The first few relations are

$$T_2 = T_0 L_2(q^2), \tag{4.23}$$

$$T_4 = T_0(3L_2^2(q^2) - 2L_4(q^2)), \tag{4.24}$$

$$T_6 = T_0(15L_2^3(q^2) - 30L_2(q^2)L_4(q^2) + 16L_6(q^2)), \tag{4.25}$$

and

$$T_8 = T_0(105L_2^4(q^2) - 420L_2^2(q^2)L_4(q^2) + 448L_2(q^2)L_6(q^2) + 140L_4^2(q^2) - 272L_8(q^2)). \tag{4.26}$$

Also note that

$$12q\frac{dT_n}{dq} = T_{n+2}.$$

Now, identities (4.23)–(4.25) imply (4.11) and (4.12). Identities (4.25) and (4.26) yield

$$-192L_2(q^2)L_6(q^2) - 80L_4^2(q^2) + 192q\frac{dL_6(q^2)}{dq} + 272L_8(q^2) = 0. \tag{4.27}$$

Using (4.21) and (4.27) and removing $L_8(q^2)$ from these identities, we deduce that

$$q\frac{dL_6(q^2)}{dq} = L_2(q^2)L_6(q^2) - L_4^2(q^2),$$

which upon substituting $x = q^2$ completes the proof of (4.13). We obtain an additional identity from this proof by eliminating $q\frac{dL_6(q^2)}{dq}$ from (4.21) and (4.27), namely,

$$L_4^2(q^2) = L_8(q^2),$$

which implies that

$$L_4^2(x) = L_8(x). \qquad \square$$

Remark 4.1. The functions S_{2k+1} and T_{2k} for non-negative integer k can be found on page 369 of Ramanujan's Lost Notebook [64]. Ramanujan indicated that S_{2k+1}/S_1 and T_{2k}/T_0 can be expressed in terms $L_2(q^2)$, $L_4(q^2)$ and $L_6(q^2)$. Ramanujan's claims were proved by Berndt and Yee [12] using Theorem 4.4. There was no indication that Ramanujan realized that his study of S_{2k+1} and T_{2k} actually leads to a proof of the more fundamental identities presented in Theorem 4.4.

Remark 4.2. Theorem 4.4 can be proved using the theory of modular forms. For more details, see [57, p. 161].

Exercises for Chapter 4

1. Let

$$I_N = \frac{1}{2\pi i} \int_{C_N} \frac{1}{\zeta^{2m+1}} \frac{2\pi i \zeta}{e^{2\pi i \zeta} - 1} \, d\zeta,$$

where C_N is the square with vertices $\pm(N + 1/2) \pm (N + 1/2)i$ traversed in the counterclockwise direction as given in the following diagram.

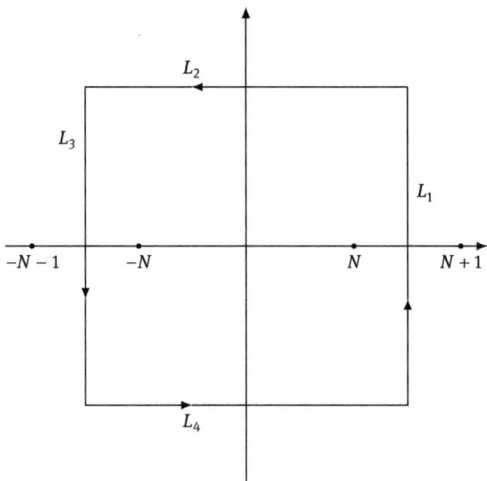

(a) Show that

$$I_N = \text{Res}(f(z); 0) + \sum_{\substack{k=-N \\ k \neq 0}}^{N} \text{Res}(f(z); k),$$

where

$$f(z) = \frac{1}{z^{2m+1}} \frac{2\pi i z}{e^{2\pi i z} - 1}.$$

(b) By giving lower bounds for $|1 - e^{2\pi i z}|$ on $L_j, j = 1, 2, 3$ and 4, use (a) to show that

$$\sum_{k=1}^{\infty} \frac{1}{k^{2j}} = \frac{(-1)^{j+1} B_{2j} (2\pi)^{2j}}{2 \cdot (2j)!}.$$

2. The Ramanujan τ function $\tau(n)$ is defined as the coefficients of the power series expansion

$$D(x) = \sum_{n=1}^{\infty} \tau(n) x^n,$$

where $D(x)$ is given by (4.14). The function $\sigma_\ell(n)$ is defined by

$$\sigma_\ell(n) = \sum_{d|n} d^\ell.$$

Show that

$$12\tau(n) = 5n\sigma_3(n) + 7n\sigma_5(n) - 2520 \sum_{j=1}^{n-1}(n-j)\sigma_3(n-j)\sigma_5(j)$$

$$+ 1680 \sum_{j=1}^{n-1}\sigma_5(n-j)\sigma_3(j).$$

Hint: Express $D(x)$ in terms of $L_4(x)$, $L_6(x)$, $x\frac{dL_6(x)}{dx}$ and $x\frac{dL_4(x)}{dx}$.

3. Given that

$$L_{10}(x) = 1 - 264 \sum_{k=1}^{\infty} \frac{k^9 x^k}{1 - x^k},$$

show that

$$L_{10}(x) = L_4(x)L_6(x)$$

by using the relations $L_8(x) = L_4^2(x)$ and the identities for T_8 and T_{10}.

Remark. The above problem shows that it can be tedious to find the explicit relation between $L_{2\ell}(x)$ and $L_{2j}(x)$, $2 \le j < \ell$. In general, relations between $L_{2\ell}(x)$ and $L_{2j}(x)$, $2 \le j < \ell$, can be derived using theory of modular forms.

5 Elliptic functions and Jacobi's triple product identity

We have seen in the previous four chapters that by using only Jacobi's triple product identity, we are able to prove many interesting identities including a system of differential equations satisfied by $L_2(x)$, $L_4(x)$ and $L_6(x)$. In this chapter, we introduce elliptic functions and give a third proof of Jacobi's triple product identity, using an elliptic function constructed from Jacobi's theta functions.

5.1 Elliptic functions

Definition 5.1. Let $\omega_1, \omega_2 \in \mathbf{C}$ be such that $\operatorname{Im} \omega_1/\omega_2 > 0$. A function f which satisfies the equations

$$f(z + \omega_1) = f(z) \quad \text{and} \quad f(z + \omega_2) = f(z)$$

for all $z \in \mathbf{C}$ for which $f(z)$ exists, is called a *doubly-periodic function* of z with periods ω_1 and ω_2. A meromorphic doubly-periodic function is called an *elliptic function*.

Definition 5.2. Let f be an elliptic function with periods ω_1 and ω_2 such that $\operatorname{Im} \omega_1/\omega_2 > 0$. The pair (ω_1, ω_2) is called a *fundamental pair* if every period of f is of the form $m\omega_1 + n\omega_2$, where m and n are integers. If (ω_1, ω_2) is a fundamental pair, then the set

$$\mathcal{F}(\omega_1, \omega_2) = \{r\omega_1 + s\omega_2 | 0 \le r < 1, 0 \le s < 1\}$$

is called a *fundamental period-parallelogram* associated with f. We use $[\omega_1, \omega_2]$ to denote the set of all linear combinations $m\omega_1 + n\omega_2$, where m and n are arbitrary integers. This is called the *lattice* generated by ω_1 and ω_2.

We will encounter several elliptic functions in the subsequent sections and chapters but first, we establish the following fundamental theorem in the theory of elliptic functions.

Theorem 5.3. *If an elliptic function f has no poles in some fundamental period-parallelogram of f, then f is a constant.*

To prove Theorem 5.3, we need the following theorem due to Liouville.

Theorem 5.4. *A bounded entire function is a constant.*

https://doi.org/10.1515/9783110541915-005

The proof of Theorem 5.4 can be found in standard textbooks on complex analysis. See for example the book by Bak and Newman [7].

Proof of Theorem 5.3. If $f(z)$ is elliptic and has no poles in a fundamental parallelogram, then since $f(z)$ is continuous, $f(z)$ is bounded on the closure of the fundamental parallelogram. By periodicity, the values of $f(z)$, $z \in \mathbf{C}$ are determined by its values on the closure of the fundamental parallelogram and hence, $f(z)$ is a bounded entire function. By Theorem 5.4, we conclude that $f(z)$ is a constant. $\square$

5.2 Transformation formulas associated with Jacobi's theta functions

In this section, we list some fundamental transformation formulas associated with $\vartheta_j(u|\tau)$, $1 \leq j \leq 4$. These will be needed for the constructions of elliptic functions in subsequent sections.

Theorem 5.5. *The following transformation formulas hold:*

$$\vartheta_1(u + \pi|\tau) = -\vartheta_1(u|\tau), \qquad\qquad \vartheta_2(u + \pi|\tau) = -\vartheta_2(u|\tau),$$

$$\vartheta_1(u + \pi\tau|\tau) = -q^{-1}e^{-2iu}\vartheta_1(u|\tau), \qquad \vartheta_2(u + \pi\tau|\tau) = q^{-1}e^{-2iu}\vartheta_2(u|\tau),$$

$$\vartheta_1\left(u + \frac{\pi}{2}\Big|\tau\right) = \vartheta_2(u|\tau), \qquad\qquad \vartheta_2\left(u + \frac{\pi}{2}\Big|\tau\right) = -\vartheta_1(u|\tau),$$

$$\vartheta_1\left(u + \frac{\pi\tau}{2}\Big|\tau\right) = iq^{-1/4}e^{-iu}\vartheta_4(u|\tau), \qquad \vartheta_2\left(u + \frac{\pi\tau}{2}\Big|\tau\right) = q^{-1/4}e^{-iu}\vartheta_3(u|\tau),$$

$$\vartheta_3(u + \pi|\tau) = \vartheta_3(u|\tau), \qquad\qquad \vartheta_4(u + \pi|\tau) = \vartheta_4(u|\tau),$$

$$\vartheta_3(u + \pi\tau|\tau) = q^{-1}e^{-2iu}\vartheta_3(u|\tau), \qquad \vartheta_4(u + \pi\tau|\tau) = -q^{-1}e^{-2iu}\vartheta_4(u|\tau),$$

$$\vartheta_3\left(u + \frac{\pi}{2}\Big|\tau\right) = \vartheta_4(u|\tau), \qquad\qquad \vartheta_4\left(u + \frac{\pi}{2}\Big|\tau\right) = \vartheta_3(u|\tau),$$

$$\vartheta_3\left(u + \frac{\pi\tau}{2}\Big|\tau\right) = q^{-1/4}e^{-iu}\vartheta_2(u|\tau), \qquad \vartheta_4\left(u + \frac{\pi\tau}{2}\Big|\tau\right) = iq^{-1/4}e^{-iu}\vartheta_1(u|\tau).$$

Proof. We will only prove four of the above transformation formulas. The proofs for the rest of the formulas are similar. Since

$$\vartheta_1(u|\tau) = -i\sum_{n=-\infty}^{\infty}(-1)^n q^{(n+1/2)^2} e^{(2n+1)iu},$$

we find that

$$\vartheta_1(u + \pi|\tau) = -i\sum_{n=-\infty}^{\infty}(-1)^n q^{(n+1/2)^2} e^{(2n+1)i(u+\pi)} = -\vartheta_1(u|\tau). \tag{5.1}$$

Next,

$$\vartheta_1(u + \pi\tau|\tau) = -i \sum_{-\infty}^{\infty} (-1)^n q^{(n+1/2)^2} e^{(2n+1)iu} e^{(2n+1)i\pi\tau}. \tag{5.2}$$

Using the fact that $q = e^{i\pi\tau}$, we rewrite (5.2) as

$$\vartheta_1(u + \pi\tau|\tau) = -i \sum_{n=-\infty}^{\infty} (-1)^n q^{n^2+3n+9/4} e^{(2(n+1)+1)iu} q^{-1} e^{-2iu}$$

$$= -q^{-1} e^{-2iu} (-i) \sum_{\ell=-\infty}^{\infty} (-1)^\ell q^{(\ell+1/2)^2} e^{(2\ell+1)iu},$$

where $\ell = n + 1$. Hence,

$$\vartheta_1(u + \pi\tau|\tau) = -q^{-1} e^{-2iu} \vartheta_1(u|\tau). \tag{5.3}$$

We now prove two transformation formulas for $\vartheta_1(u|\tau)$ involving translations of $\pi/2$ and $\pi\tau/2$. First, we observe that

$$\vartheta_1\left(u + \frac{\pi}{2}\Big|\tau\right) = -i \sum_{j=-\infty}^{\infty} (-1)^j q^{(j+1/2)^2} e^{i(2j+1)(u+\pi/2)}$$

$$= -i \sum_{j=-\infty}^{\infty} (-1)^{2j} q^{(j+1/2)^2} e^{i(2j+1)u} e^{i\pi/2}$$

$$= \vartheta_2(u|\tau).$$

Next, we find that

$$\vartheta_1\left(u + \frac{\pi\tau}{2}\Big|\tau\right) = -i \sum_{j=-\infty}^{\infty} (-1)^j q^{(j+1/2)^2} e^{i(2j+1)(u+\pi\tau/2)}$$

$$= -i \sum_{j=-\infty}^{\infty} (-1)^j q^{j^2+j+1/4} e^{i(2j+1)u} e^{i\pi/2} e^{\pi i\tau j}$$

$$= iq^{-1/4} e^{-iu} \sum_{j=-\infty}^{\infty} (-1)^{j+1} q^{j^2+2j+1} e^{2i(j+1)u}$$

$$= iq^{-1/4} e^{-iu} \vartheta_4(u|\tau). \qquad \square$$

5.3 Proof of a fundamental identity of Jacobi associated with $\vartheta_1'(0|\tau)$

We mentioned in Chapter 1 that there are several proofs of Jacobi's triple product identity (1.11) using the theory of elliptic functions. We will present one of these proofs in this section. Note that in our proof, we must take great care not to assume Theorem 3.2, which is clearly a consequence of (1.11).

We prove an important lemma about the zeros of $\vartheta_j(z|\tau)$, $1 \leq j \leq 4$. Once again, if we assume (1.11), then this lemma follows immediately from Theorem 3.2 and Theorem 5.5. Since our aim is to provide another proof of (1.11), we will prove the lemma using the argument principle from complex analysis.

Lemma 5.6. *Let C be the closed parallelogram with vertices $t, t+\pi, t+\pi\tau$ and $t+\pi+\pi\tau$ and that the boundary ∂C of C does not contain any zeroes of $\vartheta_j(u|\tau), 1 \leq j \leq 4$. Then $\vartheta_j(u|\tau)$ has exactly one zero inside C.*

Proof. In view of the transformation formulas of $\vartheta_j(z|\tau)$, $1 \leq j \leq 4$, it suffices to consider the zero of $\vartheta_1(z|\tau)$. From the theory of complex analysis, we know that the number of zero of $\vartheta_1(z|\tau)$ in C is given by

$$\frac{1}{2\pi i} \int_{\partial C} \frac{\vartheta_1'(s|\tau)}{\vartheta_1(s|\tau)} \, ds,$$

where ∂C is the boundary of C traversed in the counterclockwise direction. By direct computations, we deduce that

$$\frac{1}{2\pi i} \int_{\partial C} \frac{\vartheta_1'(s|\tau)}{\vartheta_1(s|\tau)} \, ds = \frac{1}{2\pi i} \int_{t}^{t+\pi} \frac{\vartheta_1'(s|\tau)}{\vartheta_1(s|\tau)} \, ds + \frac{1}{2\pi i} \int_{t+\pi}^{t+\pi+\pi\tau} \frac{\vartheta_1'(s|\tau)}{\vartheta_1(s|\tau)} \, ds$$

$$+ \frac{1}{2\pi i} \int_{t+\pi+\pi\tau}^{t+\pi\tau} \frac{\vartheta_1'(s|\tau)}{\vartheta_1(s|\tau)} \, ds + \frac{1}{2\pi i} \int_{t+\pi\tau}^{t} \frac{\vartheta_1'(s|\tau)}{\vartheta_1(s|\tau)} \, ds$$

$$= \frac{1}{2\pi i} \int_{t}^{t+\pi} \frac{\vartheta_1'(s|\tau)}{\vartheta_1(s|\tau)} \, ds - \frac{1}{2\pi i} \int_{t}^{t+\pi} \frac{\vartheta_1'(s+\pi\tau|\tau)}{\vartheta_1(s+\pi\tau|\tau)} \, ds$$

$$- \frac{1}{2\pi i} \int_{t}^{t+\pi\tau} \frac{\vartheta_1'(s|\tau)}{\vartheta_1(s|\tau)} \, ds + \frac{1}{2\pi i} \int_{t}^{t+\pi\tau} \frac{\vartheta_1'(s+\pi|\tau)}{\vartheta_1(s+\pi|\tau)} \, ds$$

$$= 1,$$

where the last equality follows from the transformation formulas

$$\frac{\vartheta_1'(s+\pi|\tau)}{\vartheta_1(s+\pi|\tau)} = \frac{\vartheta_1'(s|\tau)}{\vartheta_1(s|\tau)}$$

and

$$\frac{\vartheta_1'(s+\pi\tau|\tau)}{\vartheta_1(s+\pi\tau|\tau)} = \frac{\vartheta_1'(s|\tau)}{\vartheta_1(s|\tau)} - 2i.$$

This completes the proof of the lemma. □

Corollary 5.7. *Let* $z_1 = 0$, $z_2 = \pi/2$, $z_3 = \pi(\tau+1)/2$ *and* $z_4 = \pi\tau/2$. *Then, for* $j = 1, 2, 3$ *and* 4, $\vartheta_j(u|\tau)$ *has a simple zero at* z_j *in* $\mathcal{F}(\pi\tau, \pi)$.

Proof. By the definition of $\vartheta_1(u|\tau)$ (see (3.1)) and Lemma 5.6, we conclude that $\vartheta_1(u|\tau)$ has a simple zero at $z_1 = 0$ in some closed parallelogram C containing 0 with vertices $t, t + \pi, t + \pi\tau$ and $t + \pi(\tau + 1)$ such that 0 is not on the boundary. This shows that 0 is the only simple zero of $\vartheta_1(u|\tau)$ in $\mathcal{F}(\pi\tau, \pi)$.

By Theorem 5.5, we conclude that the simple zeroes of $\vartheta_j(u|\tau)$ are $\pi/2, \pi(\tau+1)/2$ and $\pi\tau/2$, respectively for $j = 2, 3$ and 4. This completes the proof of the corollary. $\square$

The key to our proof of (1.11) using elliptic functions is the following identity.

Theorem 5.8. *Let* τ *be complex numbers with* $\operatorname{Im}\tau > 0$. *Then*

$$\vartheta_1'(0|\tau) = \vartheta_2(0|\tau)\vartheta_3(0|\tau)\vartheta_4(0|\tau). \tag{5.4}$$

Identity (5.4) is a consequence of (1.11) (see Exercise 3 of Chapter 3). However, since our task is to prove (1.11), we need to derive (5.4) using an approach that is independent of (1.11).

Proof. Let

$$A(u|\tau) = \frac{\vartheta_1(2u|2\tau)}{\vartheta_2(u|\tau)\vartheta_1(u|\tau)}.$$

Using Theorem 5.5, we deduce that $A(u + \pi|\tau) = A(u|\tau)$ and $A(u + \pi\tau|\tau) = A(u)$ and we conclude that $A(u|\tau)$ is an elliptic function with periods π and $\pi\tau$. Let $\alpha \in \mathbf{C}$ be such that the parallelogram C with vertices α, $\alpha + \pi$, $\alpha + \pi\tau$ and $\alpha + \pi + \pi\tau$ contains 0 and $\pi/2$. Observe that by Corollary 5.7, 0 and $\pi/2$ are the only simple zeroes of the denominator of $A(u|\tau)$ in C. On the other hand, 0 and $\pi/2$ are zeroes of the numerator of $A(u|\tau)$. Hence, $A(u|\tau)$ is an entire elliptic function and by Theorem 5.3, $A(u|\tau)$ must be a constant. Let

$$c = \frac{\vartheta_1(2u|2\tau)}{\vartheta_2(u|\tau)\vartheta_1(u|\tau)}.$$

Replacing u by $u + \pi\tau/2$, we find that

$$c = \frac{\vartheta_4(2u|2\tau)}{\vartheta_3(u|\tau)\vartheta_4(u|\tau)}.$$

By setting $u = 0$, we conclude that

$$c = \frac{\vartheta_4(0|2\tau)}{\vartheta_3(0|\tau)\vartheta_4(0|\tau)}.$$

Therefore, we must have

$$\vartheta_3(0|\tau)\vartheta_4(0|\tau)\vartheta_1(2u|2\tau) = \vartheta_2(u|\tau)\vartheta_1(u|\tau)\vartheta_4(0|2\tau). \tag{5.5}$$

Differentiating both sides of (5.5) with respect to u and setting $u = 0$, we deduce that

$$2\vartheta_1'(0|2\tau)\vartheta_3(0|\tau)\vartheta_4(0|\tau) = \vartheta_4(0|2\tau)\vartheta_2(0|\tau)\vartheta_1'(0|\tau).$$

In other words,

$$2\frac{\vartheta_1'(0|2\tau)}{\vartheta_4(0|2\tau)\vartheta_2^2(0|\tau)} = \frac{\vartheta_1'(0|\tau)}{\vartheta_2(0|\tau)\vartheta_3(0|\tau)\vartheta_4(0|\tau)}. \tag{5.6}$$

We next show that

$$2\vartheta_2(0|2\tau)\vartheta_3(0|2\tau) = \vartheta_2^2(0|\tau). \tag{5.7}$$

Our proof of (5.7) is similar to the proof of Theorem 2.1. Consider the identity

$$2\vartheta_2(0|4\tau)\vartheta_3(0|4\tau) = \left(\vartheta_2(0|4\tau) + \vartheta_3(0|4\tau)\right)^2 - \vartheta_2^2(0|4\tau) - \vartheta_3^2(0|4\tau). \tag{5.8}$$

Note that

$$\vartheta_2(0|4\tau) + \vartheta_3(0|4\tau) = \vartheta_3(0|\tau)$$

and we may simplify (5.8) as

$$2\vartheta_2(0|4\tau)\vartheta_3(0|4\tau) = \vartheta_3^2(0|\tau) - \vartheta_2^2(0|4\tau) - \vartheta_3^2(0|4\tau). \tag{5.9}$$

Observe that

$$\vartheta_2^2(0|4\tau) + \vartheta_3^2(0|4\tau) = \sum_{\ell^\dagger \in L^\dagger} q^{N(\ell^\dagger)} \tag{5.10}$$

and

$$\vartheta_3^2(0|\tau) = \sum_{\ell^\dagger \in L^\dagger} q^{N(\ell^\dagger)} + \sum_{\ell^\dagger \in L^\dagger} q^{N(\ell^\dagger+1)}, \tag{5.11}$$

where $L^\dagger$ is defined in (2.3). Substituting (5.10) and (5.11) in (5.9), we deduce that

$$2\vartheta_2(0|4\tau)\vartheta_3(0|4\tau) = \sum_{n,m \in \mathbf{Z}} q^{(m+n+1)^2+(m-n)^2} = \vartheta_2^2(0|2\tau).$$

Replacing 2τ by τ, we deduce (5.7).

Substituting (5.7) into (5.6), we find that

$$\frac{\vartheta_1'(0|\tau)}{\vartheta_2(0|\tau)\vartheta_3(0|\tau)\vartheta_4(0|\tau)} = \frac{\vartheta_1'(0|2\tau)}{\vartheta_2(0|2\tau)\vartheta_3(0|2\tau)\vartheta_4(0|2\tau)}. \tag{5.12}$$

Iterating the above, we conclude that

$$\frac{\vartheta_1'(0|\tau)}{\vartheta_2(0|\tau)\vartheta_3(0|\tau)\vartheta_4(0|\tau)} = \frac{\vartheta_1'(0|2^n\tau)}{\vartheta_2(0|2^n\tau)\vartheta_3(0|2^n\tau)\vartheta_4(0|2^n\tau)}.$$

As n tends to infinity, the right-hand side is 1 and so, we complete the proof of (5.4). $\qquad\square$

Remark 5.1. Identity (5.4) was discovered and proved by Jacobi in [52, pp. 515–517]. His proof involved establishing

$$\frac{\vartheta_1'(0|\tau)}{\vartheta_2(0|\tau)\vartheta_3(0|\tau)\vartheta_4(0|\tau)} = \frac{\vartheta_1'(0|4\tau)}{\vartheta_2(0|4\tau)\vartheta_3(0|4\tau)\vartheta_4(0|4\tau)} \tag{5.13}$$

using theta functions of three variables. For a presentation of Jacobi's proof of (5.4), see Lawden's book [58, Section 1.5]. Our proof avoids the use of identities involving more than two variables and we prove (1.11) using (5.12) instead of (5.13). There are at least two other proofs of (5.4) (which of course translates to two other proofs of (1.11)) that use the theory of elliptic functions. For more details, see [72, pp. 469–473] and [72, p. 490].

Remark 5.2. The identity (5.5) used to derive (5.4) is due to Landen and it is usually referred to as Landen's transformation formula for theta functions [13, Theorem 2.5].

5.4 Proof of the Jacobi triple product identity using theory of elliptic functions

We are now ready to prove

$$\sum_{j=-\infty}^{\infty} q^{j^2} e^{2iuj} = \prod_{j=1}^{\infty} (1 - q^{2j})(1 + e^{2iu} q^{2j-1})(1 + e^{-2iu} q^{2j-1}), \tag{5.14}$$

which is equivalent to (1.11).

Let

$$f(u) = \vartheta_3(u|\tau)$$

and

$$P(u|\tau) = \prod_{j=1}^{\infty}(1 + e^{2iu}q^{2j-1})(1 + e^{-2iu}q^{2j-1}).$$

Observe that both $\vartheta_3(u|\tau)$ and $P(u|\tau)$ satisfy the functional equations

$$F(z + \pi|\tau) = F(z|\tau)$$

and

$$F(z + \pi\tau|\tau) = q^{-1}e^{-2iu}F(z|\tau).$$

Now, in $\mathcal{F}(\pi\tau, \pi)$, the only simple zero of $\vartheta_3(u|\tau)$ is $\pi(\tau + 1)/2$ by Corollary 5.7. This is also true for the function $P(u|\tau)$. This implies that $\vartheta_3(u|\tau)/P(u|\tau)$ is an elliptic function with periods π and $\pi\tau$ with no pole in $\mathcal{F}(\pi\tau, \pi)$. By Theorem 5.3,

$$\vartheta_3(u|\tau) = G(q) \prod_{j=1}^{\infty}(1 + e^{2iu}q^{2j-1})(1 + e^{-2iu}q^{2j-1}) \qquad (5.15)$$

for some function $G(q)$ independent of u.

Using (5.15) and Theorem 5.5, we deduce that

$$\vartheta_1(u|\tau) = -i \sum_{j=-\infty}^{\infty} (-1)^j q^{(j+\frac{1}{2})^2} e^{(2j+1)iu}$$

$$= 2q^{1/4} G(q) \sin u \prod_{j=1}^{\infty}(1 - e^{2iu}q^{2j})(1 - e^{-2iu}q^{2j}),$$

$$\vartheta_2(u|\tau) = \sum_{j=-\infty}^{\infty} q^{(j+\frac{1}{2})^2} e^{(2j+1)iu}$$

$$= 2q^{1/4} G(q) \cos u \prod_{j=1}^{\infty}(1 + e^{2iu}q^{2j})(1 + e^{-2iu}q^{2j}),$$

and

$$\vartheta_4(u|\tau) = \sum_{j=-\infty}^{\infty} (-1)^j q^{j^2} e^{2jiu}$$

$$= G(q) \prod_{j=1}^{\infty}(1 - e^{2iu}q^{2j-1})(1 - e^{-2iu}q^{2j-1}).$$

This implies that

$$\vartheta_1'(0|\tau) = 2q^{1/4} G(q) \prod_{j=1}^{\infty}(1 - q^{2j})^2, \qquad (5.16)$$

$$\vartheta_2(0|\tau) = 2q^{1/4} G(q) \prod_{j=1}^{\infty} (1 + q^{2j})^2, \tag{5.17}$$

$$\vartheta_3(0|\tau) = G(q) \prod_{j=1}^{\infty} (1 + q^{2j-1})^2, \tag{5.18}$$

and

$$\vartheta_4(0|\tau) = G(q) \prod_{j=1}^{\infty} (1 - q^{2j-1})^2. \tag{5.19}$$

Substituting (5.16)–(5.19) into (5.4), we deduce that

$$G(q) \prod_{j=1}^{\infty} (1 - q^{2j})^2 = G^3(q) \prod_{j=1}^{\infty} (1 - q^{2j-1})^2 (1 + q^{2j-1})^2 (1 + q^{2j})^2 = G^3(q),$$

where we have used the fact that

$$\prod_{j=1}^{\infty} (1 - q^{2j-1})^2 (1 + q^{2j-1})^2 (1 + q^{2j})^2 = 1.$$

This implies that

$$G(q) = \pm \prod_{j=1}^{\infty} (1 - q^{2j}).$$

The "plus" sign is chosen after comparing the constant coefficients of both sides of (5.15) and the proof of (5.14) is complete.

Exercises for Chapter 5

1. Let τ_1 and τ_2 be two complex numbers with $\tau_1/\tau_2 \notin \mathbf{R}$ and let $[\tau_1, \tau_2]$ be given as in Definition 5.2. Given two pairs of complex numbers (ω_1, ω_2) and (ω_1', ω_2') with nonreal ratios ω_2/ω_1 and ω_2'/ω_1'. Prove that

$$[\omega_1, \omega_2] = [\omega_1', \omega_2']$$

if and only if there is a 2×2 matrix $\left(\begin{smallmatrix} a & b \\ c & d \end{smallmatrix}\right)$ with integer entries and determinant ± 1 such that

$$\begin{pmatrix} \omega_1' \\ \omega_2' \end{pmatrix} = \begin{pmatrix} a & b \\ c & d \end{pmatrix} \begin{pmatrix} \omega_1 \\ \omega_2 \end{pmatrix}.$$

2. Let $f(z)$ be a non-constant elliptic function with periods ω_1 and ω_2. Show that $f(z)$ must have at least two simple poles, or at least one pole which is not simple, in the parallelogram with vertices $\alpha, \alpha + \omega_1, \alpha + \omega_2$, and $\alpha + \omega_1 + \omega_2$ for any $\alpha \in \mathbf{C}$.

3. Let Z denote the sum of the zeroes of an elliptic function f in a fundamental period-parallelogram of f and let P denote the sum of the poles in the same parallelogram. Show that $Z - P$ is a period of f. (Hint: Integrate $zf'(z)/f(z)$.)

4. (a) Show that the function

$$\frac{2\vartheta_1(u|\tau)\vartheta_2(u|\tau)\vartheta_3(u|\tau)\vartheta_4(u|\tau)}{\vartheta_1(2u|\tau)\vartheta_2(0|\tau)\vartheta_3(0|\tau)\vartheta_4(0|\tau)}$$

   is a constant.

   (b) By comparing coefficients of u^2 in the expansions of

$$\ln \frac{\vartheta_1(2u|\tau)}{2\vartheta_1(u|\tau)}$$

   and

$$\ln \frac{\vartheta_2(u|\tau)}{\vartheta_2(0|\tau)} + \ln \frac{\vartheta_3(u|\tau)}{\vartheta_3(0|\tau)} + \ln \frac{\vartheta_4(u|\tau)}{\vartheta_4(0|\tau)},$$

   deduce that

$$\frac{\vartheta_1'''(0|\tau)}{\vartheta_1'(0|\tau)} = \frac{\vartheta_2''(0|\tau)}{\vartheta_2(0|\tau)} + \frac{\vartheta_3''(0|\tau)}{\vartheta_3(0|\tau)} + \frac{\vartheta_4''(0|\tau)}{\vartheta_4(0|\tau)}. \tag{5.20}$$

   (c) Show that (5.20) implies

$$\vartheta_1'(0|\tau) = \vartheta_2(0|\tau)\vartheta_3(0|\tau)\vartheta_4(0|\tau).$$

6 Two elliptic functions and their properties

In this chapter, we introduce Weierstrass' $\wp$ function and derive a differential equation associated with this function. We then introduce another elliptic function arising from the study of the product representation of $\vartheta_1(u|\tau)$ and show the connection between Lambert's series $L_{2j}(x)$ and Eisenstein's series $E_{2j}(\tau)$. Eisenstein's series satisfy certain transformation formulas and this implies that $L_{2j}(e^{2\pi i \tau})$, $j \geq 2$, satisfy the same transformation formulas. Together with (4.15), we derive a transformation formula satisfied by Dedekind's function $\Delta(\tau)$.

6.1 Weierstrass' $\wp$ function

Let $\omega_1, \omega_2 \in \mathbf{C}$ with $\omega_1/\omega_2 \notin \mathbf{R}$ and

$$[\omega_1, \omega_2] = \{m\omega_1 + n\omega_2 | m, n \in \mathbf{Z}\}.$$

Definition 6.1. The Weierstrass $\wp$ function is defined by the series

$$\wp(u|[\omega_1, \omega_2]) = \frac{1}{u^2} + \sum_{\substack{\ell \in [\omega_1, \omega_2] \\ \ell \neq 0}} \left(\frac{1}{(u-\ell)^2} - \frac{1}{\ell^2} \right). \tag{6.1}$$

The Weierstrass $\wp$ function is analytic except for a double pole at points in $[\omega_1, \omega_2]$. For proofs of these properties of $\wp$, see Apostol's book [5, Theorem 1.10]. We will prove that

Theorem 6.2. *The Weierstrass $\wp$ is an even elliptic function with periods ω_1 and ω_2.*

Proof. We first observe that

$$\wp(u + \omega_1|[\omega_1, \omega_2]) = \frac{1}{(u+\omega_1)^2} + \sum_{\substack{\ell \in [\omega_1, \omega_2] \\ \ell \neq 0}} \left(\frac{1}{(u+\omega_1-\ell)^2} - \frac{1}{\ell^2} \right)$$

$$= \frac{1}{(u+\omega_1)^2} + \sum_{\substack{\ell \in [\omega_1, \omega_2] \\ \ell \neq 0 \\ \ell \neq \omega_1}} \left(\frac{1}{(u+\omega_1-\ell)^2} - \frac{1}{\ell^2} \right) + \frac{1}{u^2} - \frac{1}{\omega_1^2}$$

$$= \frac{1}{(u+\omega_1)^2} + \sum_{\substack{\ell \in [\omega_1, \omega_2] \\ \ell \neq 0 \\ \ell \neq \omega_1}} \left(\frac{1}{(u+\omega_1-\ell)^2} - \frac{1}{(\ell-\omega_1)^2} \right)$$

$$\qquad - \sum_{\substack{\ell \in [\omega_1, \omega_2] \\ \ell \neq 0 \\ \ell \neq \omega_1}} \left(\frac{1}{\ell^2} - \frac{1}{(\ell-\omega_1)^2} \right) + \frac{1}{u^2} - \frac{1}{\omega_1^2}. \tag{6.2}$$

https://doi.org/10.1515/9783110541915-006

Now,

$$\sum_{\substack{\ell \in [\omega_1, \omega_2] \\ \ell \neq 0 \\ \ell \neq \omega_1}} \left(\frac{1}{\ell^2} - \frac{1}{(\ell - \omega_1)^2} \right) = \sum_{\substack{\ell' = \omega_1 - \ell \in [\omega_1, \omega_2] \\ \ell' \neq 0 \\ \ell' \neq \omega_1}} \left(\frac{1}{(\omega_1 - \ell')^2} - \frac{1}{(-\ell')^2} \right)$$

$$= - \sum_{\substack{\ell \in [\omega_1, \omega_2] \\ \ell \neq 0 \\ \ell \neq \omega_1}} \left(\frac{1}{\ell^2} - \frac{1}{(\ell - \omega_1)^2} \right).$$

This implies that

$$\sum_{\substack{\ell \in [\omega_1, \omega_2] \\ \ell \neq 0 \\ \ell \neq \omega_1}} \left(\frac{1}{\ell^2} - \frac{1}{(\ell - \omega_1)^2} \right) = 0.$$

We therefore conclude from (6.2) that

$$\wp(u + \omega_1 | [\omega_1, \omega_2]) = \wp(u | [\omega_1, \omega_2]).$$

In a similar way, we deduce that

$$\wp(u + \omega_2 | [\omega_1, \omega_2]) = \wp(u | [\omega_1, \omega_2]).$$

The fact that $\wp$ is even follows from the identity

$$(-u - \ell)^2 = (u + \ell)^2 = (u - (-\ell))^2$$

and the fact that the map that sends ℓ to $-\ell$ is a bijection on $[\omega_1, \omega_2]$. □

6.2 A differential equation satisfied by Weierstrass' $\wp$ function

Definition 6.3. Let $k \geq 2$. Suppose $\omega_1, \omega_2 \in \mathbf{C}$ with $\omega_1/\omega_2 \notin \mathbf{R}$. We define

$$\mathcal{E}_{2k}([\omega_1, \omega_2]) = \sum_{\substack{\ell \in [\omega_1, \omega_2] \\ \ell \neq 0}} \frac{1}{\ell^{2k}}. \tag{6.3}$$

If we expand $\wp(u | [\omega_1, \omega_2])$ as a power series about 0, then

$$\wp(u | [\omega_1, \omega_2]) = \frac{1}{u^2} + \sum_{\substack{\ell \in [\omega_1, \omega_2] \\ \ell \neq 0}} \frac{1}{\ell^2} \left(\sum_{m=1}^{\infty} m \left(\frac{u}{\ell} \right)^{m-1} - 1 \right)$$

$$= \frac{1}{u^2} + \sum_{m=2}^{\infty} m \sum_{\substack{\ell \in [\omega_1, \omega_2] \\ \ell \neq 0}} \frac{u^{m-1}}{\ell^{m-1}}. \tag{6.4}$$

Since $\wp$ is an even function in u, we may let $m = 2k + 1$ and deduce from (6.4) that

$$\wp(u|[\omega_1, \omega_2]) = \frac{1}{u^2} + \sum_{k=1}^{\infty} (2k + 1)\mathcal{E}_{2k+2}([\omega_1, \omega_2])u^{2k} \tag{6.5}$$

$$= \frac{1}{u^2} + 3\mathcal{E}_4([\omega_1, \omega_2])u^2 + 5\mathcal{E}_6([\omega_1, \omega_2])u^4 + \cdots \tag{6.6}$$

where $\mathcal{E}_{2k+2}([\omega_1, \omega_2])$ is given by (6.3). Next we differentiate both sides of (6.6) with respect to u to deduce that

$$\wp'(u|[\omega_1, \omega_2]) = -\frac{2}{u^3} + 6\mathcal{E}_4([\omega_1, \omega_2])u + 20\mathcal{E}_6([\omega_1, \omega_2])u^3 + \cdots. \tag{6.7}$$

Now, $\wp'(u|[\omega_1, \omega_2])$ is an elliptic function with periods ω_1 and ω_2 and the function $(\wp'(u|[\omega_1, \omega_2]))^2$ has a pole of order 6 at $u = 0$. Using (6.6) and (6.7), we deduce that

$$(\wp'(u|[\omega_1, \omega_2]))^2 - 4\wp^3(u|[\omega_1, \omega_2]) = -\frac{60\mathcal{E}_4([\omega_1, \omega_2])}{u^2} - 140\mathcal{E}_6([\omega_1, \omega_2]) + \cdots.$$

This implies that

$$(\wp'(u|[\omega_1, \omega_2]))^2 - 4\wp^3(u|[\omega_1, \omega_2]) + 60\mathcal{E}_4([\omega_1, \omega_2])\wp(u|[\omega_1, \omega_2])$$

is an elliptic function without any pole. In other words, it must be a constant given by $-140\mathcal{E}_6([\omega_1, \omega_2])$. This leads to the following theorem.

Theorem 6.4. *The Weierstrass $\wp$ function satisfies the differential equation*

$$(\wp'(u|[\omega_1, \omega_2]))^2 = \wp^3(u|[\omega_1, \omega_2]) - 60\mathcal{E}_4([\omega_1, \omega_2])\wp(u|[\omega_1, \omega_2])$$
$$- 140\mathcal{E}_6([\omega_1, \omega_2]).$$

6.3 An elliptic function $J(u|\tau)$ associated with $\vartheta_1(u|\tau)$

In this section, we construct another elliptic function from $\vartheta_1(u|\tau)$.

Definition 6.5. Let $u, \tau \in \mathbf{C}$ with $\operatorname{Im} \tau > 0$ and define

$$J(u|\tau) = \left(\frac{\vartheta_1'(u|\tau)}{\vartheta_1(u|\tau)} \right)'.$$

Theorem 6.6. *The function $J(u|\tau)$ is an elliptic function with periods π and $\pi\tau$.*

Proof. Now, from (5.1) and (5.3), we deduce that

$$\vartheta_1'(u + \pi|\tau) = -\vartheta_1'(u|\tau) \tag{6.8}$$

and

$$\vartheta_1'(u + \pi\tau|\tau) = -q^{-1}\vartheta_1'(u|\tau)e^{-2iu} + 2ie^{-2iu}q^{-1}\vartheta_1(u|\tau). \tag{6.9}$$

Using (5.1), (5.3), (6.8) and (6.9), we conclude that

$$\frac{\vartheta_1'(u + \pi|\tau)}{\vartheta_1(u + \pi|\tau)} = \frac{\vartheta_1'(u|\tau)}{\vartheta_1(u|\tau)}. \tag{6.10}$$

Note that

$$\begin{aligned}
\frac{\vartheta_1'(u + \pi\tau|\tau)}{\vartheta_1(u + \pi\tau|\tau)} &= -q^{-1}e^{-2iu}\frac{\vartheta_1'(u|\tau)}{\vartheta_1(u + \pi\tau|\tau)} + 2ie^{-2iu}q^{-1}\frac{\vartheta_1(u|\tau)}{\vartheta_1(u + \pi\tau|\tau)} \\
&= \frac{-q^{-1}e^{-2iu}}{-e^{-2iu}q^{-1}} \cdot \frac{\vartheta_1'(u|\tau)}{\vartheta_1(u|\tau)} - \frac{2ie^{-2iu}q^{-1}}{e^{-2iu}q^{-1}} \\
&= \frac{\vartheta_1'(u|\tau)}{\vartheta_1(u|\tau)} - 2i.
\end{aligned} \tag{6.11}$$

Differentiating (6.10) and (6.11) with respect to u, we deduce that the function $\left(\frac{\vartheta_1'(u|\tau)}{\vartheta_1(u|\tau)}\right)'$ is an elliptic function with periods π and $\pi\tau$. □

6.4 Weierstrass' $\wp$ functions and an elliptic function associated with $\vartheta_1(u|\tau)$

Let

$$J(u|\tau) = \left(\frac{\vartheta_1'(u|\tau)}{\vartheta_1(u|\tau)}\right)'. \tag{6.12}$$

Theorem 6.6 shows that $J(u|\tau)$ is an elliptic function with periods π and $\pi\tau$. Now, from Lemma 5.6, we know that the only zero of $\vartheta_1(u|\tau)$ in $\mathcal{F}(\pi\tau, \pi)$ is 0 and this is a simple zero. Since $\vartheta_1(u|\tau)$ is analytic everywhere, the function $\vartheta_1'(u|\tau)/\vartheta_1(u|\tau)$ has a simple pole at $u = 0$. Therefore, $J(u|\tau)$ has a pole of order 2 at $u = 0$ in $\mathcal{F}(\pi\tau, \pi)$. From (4.7), we find that $J(u|\tau)$ is an even function. Summarizing, we conclude that $J(u|\tau)$ is an even elliptic function with a pole of order 2 at $u = 0$. Since $\wp(u|[\pi\tau, \pi])$

is also an even elliptic function with a pole of order 2 at $u = 0$ in $\mathcal{F}(\pi\tau, \pi)$, we deduce that, for some constant A,

$$J(u|\tau) - A\wp(u|[\pi\tau, \pi])$$

is an entire elliptic function and by Theorem 5.3, this difference must be a constant. In other words, we have

$$\left(\frac{\vartheta_1'(u|\tau)}{\vartheta_1(u|\tau)}\right)' = A\wp(u|[\pi\tau, \pi]) + B \tag{6.13}$$

where A and B are some constants independent of u.

Now, by differentiating both sides of (4.9) with respect to u, we deduce that

$$\left(\frac{\vartheta_1'(u|\tau)}{\vartheta_1(u|\tau)}\right)' = -\frac{1}{u^2} - \frac{L_2(e^{2\pi i\tau})}{3}$$
$$+ \sum_{k=1}^{\infty}(-1)^{k+1}\frac{2^{2k+2}B_{2k+2}L_{2k+2}(e^{2\pi i\tau})}{(2k+2)(2k)!}u^{2k}, \tag{6.14}$$

where $B_{2j}, j = 2, 3, \ldots,$ are the Bernoulli numbers and the $L_{2j}(x)$ (see (4.8)) are given by

$$L_2(x) = 1 - 24\sum_{\ell=1}^{\infty}\frac{\ell x^\ell}{1 - x^\ell}$$

and

$$L_{2j}(x) = 1 - \frac{4j}{B_{2j}}\sum_{\ell=1}^{\infty}\frac{\ell^{2j-1}x^\ell}{1 - x^\ell}, \quad j \geq 2.$$

Using (6.5) and (6.14), we determine A and B in (6.13) and deduce the following identity.

Theorem 6.7. *Let $u \in \mathbf{C}$ and* $\mathrm{Im}\,\tau > 0$. *We have*

$$\left(\frac{\vartheta_1'(u|\tau)}{\vartheta_1(u|\tau)}\right)' = -\wp(u|[\pi\tau, \pi]) - \frac{L_2(e^{2\pi i\tau})}{3}. \tag{6.15}$$

6.5 Transformation formula satisfied by Dedekind's function $\Delta(\tau)$

In this section, we deduce a transformation formula satisfied by $\Delta(\tau)$ using (4.15) and Theorem 6.7.

Using (6.14) and (6.5), we deduce from (6.15) that

$$-\frac{1}{u^2} - \frac{L_2(e^{2\pi i \tau})}{3} + \sum_{k=1}^{\infty}(-1)^{k+1}\frac{2^{2k+2}B_{2k+2}L_{2k+2}(e^{2\pi i \tau})}{(2k+2)(2k)!}u^{2k}$$

$$= -\frac{1}{u^2} - \sum_{k=1}^{\infty}(2k+1)\mathcal{E}_{2k+2}([\pi\tau,\pi])u^{2k} - \frac{L_2(e^{2\pi i \tau})}{3}. \tag{6.16}$$

By comparing coefficients of u^{2k-2} in (6.16) and using the definition of $\mathcal{E}_{2k}([\omega_1,\omega_2])$ in (6.3), we deduce that, for $k \geq 2$,

$$\mathcal{E}_{2k}([\pi\tau,\pi]) = \sum_{\substack{\ell \in [\pi\tau,\pi] \\ \ell \neq (0,0)}} \frac{1}{\ell^{2k}} = \frac{(-1)^{k-1}2^{2k}}{(2k)!}B_{2k}L_{2k}(e^{2\pi i \tau}). \tag{6.17}$$

Definition 6.8. For $j \geq 2$, let

$$E_{2j}(\tau) = \frac{(2j)!}{(-1)^{k-1}2^{2k}B_{2k}}\mathcal{E}_{2j}([\pi\tau,\pi])$$

The functions $E_{2j}(\tau)$ are called *Eisenstein's series* of weight $2j$.

Identity (6.17) shows that

$$E_{2j}(\tau) = L_{2j}(e^{2\pi i \tau}). \tag{6.18}$$

In other words, $L_{2j}(e^{2\pi i \tau})$ is the Fourier series expansion of Eisenstein's series $E_{2j}(\tau)$.

Let $a, b, c, d \in \mathbf{Z}$ with $ad - bc = 1$. Then

$$[\pi\tau,\pi] = [\pi(a\tau + b), \pi(c\tau + d)].$$

This implies that, for $j \geq 2$,

$$\mathcal{E}_{2j}([\pi\tau,\pi]) = \mathcal{E}_{2j}([\pi(a\tau + b), \pi(c\tau + d)])$$

$$= \sum_{\substack{\ell' \in [\pi(a\tau+b),\pi(c\tau+d)] \\ \ell \neq (0,0)}} \frac{1}{\ell'^{2j}}$$

$$= \frac{1}{(c\tau + d)^{2j}}\mathcal{E}_{2j}\left(\left[\pi\frac{a\tau + b}{c\tau + d},\pi\right]\right). \tag{6.19}$$

From (6.19), we thus obtain the following theorem.

Theorem 6.9. *Let $j \geq 2$ be a positive integer and $E_{2j}(\tau)$ be given by (6.18). For integers a, b, c, d satisfying $ad - bc = 1$,*

$$E_{2j}\left(\frac{a\tau + b}{c\tau + d}\right) = (c\tau + d)^{2j}E_{2j}(\tau). \tag{6.20}$$

Applying (6.20) to the cases when $j = 2$ and 3, we conclude that

$$E_4\left(\frac{a\tau + b}{c\tau + d}\right) = (c\tau + d)^4 E_4(\tau) \tag{6.21}$$

and

$$E_6\left(\frac{a\tau + b}{c\tau + d}\right) = (c\tau + d)^6 E_6(\tau). \tag{6.22}$$

Recall from Theorem 4.5 that

$$D(x) = x \prod_{k=1}^{\infty} (1 - x^k)^{24} = \frac{1}{1728}(L_4^3(x) - L_6^2(x)). \tag{6.23}$$

Let

$$\Delta(\tau) = D(e^{2\pi i \tau}). \tag{6.24}$$

Then, by (6.24), (6.23) and (6.18), we deduce that

$$\Delta(\tau) = \frac{1}{1728}(E_4^3(\tau) - E_6^2(\tau)). \tag{6.25}$$

Using the representation (6.25), the transformation formulas (6.21) and (6.22), we deduce the following theorem.

Theorem 6.10. *For $\tau \in \mathbf{C}$ with $\operatorname{Im} \tau > 0$ and $a, b, c, d \in \mathbf{Z}$ such that $ad - bc = 1$,*

$$\Delta\left(\frac{a\tau + b}{c\tau + d}\right) = (c\tau + d)^{12}\Delta(\tau), \tag{6.26}$$

where $\Delta(\tau)$ is given by (6.25).

Remark 6.1. It is possible to show (6.26) without using Theorem 4.5. See, for example, the article by Siegel [68], Apostol's book [5, pp. 47–51] or the last chapter of Serre's book [67, pp. 95–96].

Remark 6.2. Some authors (see for example Apostol's book [5, p. 24]) refer to $\mathcal{E}_{2j}([\pi\tau, \pi])$ as Eisenstein's series and $E_{2j}(\tau)$ as normalized Eisenstein series.

Definition 6.11. The *Dedekind η function* is defined by

$$\eta(\tau) = e^{\pi i \tau / 12} \prod_{k=1}^{\infty} (1 - e^{2\pi i k \tau}). \tag{6.27}$$

From (6.26), we deduce that

$$\eta(-1/\tau) = \epsilon\sqrt{\tau}\eta(\tau) \tag{6.28}$$

for some 24th root of unity ϵ. When $\tau = i$, we conclude from (6.28) that $\epsilon = \sqrt{-i}$. We record the final result as follows.

Theorem 6.12. *If $\tau \in \mathbf{C}$ and $\operatorname{Im}\tau > 0$, then*

$$\eta(-1/\tau) = \sqrt{-i\tau}\eta(\tau). \tag{6.29}$$

A consequence of Theorem 6.12 is the following transformation formula for $\vartheta_3(0|\tau)$.

Theorem 6.13. *If $\tau \in \mathbf{C}$ and $\operatorname{Im}\tau > 0$, then*

$$\vartheta_3(0\,|-1/\tau) = \sqrt{-i\tau}\vartheta_3(0|\tau). \tag{6.30}$$

Proof. Recall from (3.4) that

$$\begin{aligned}
\vartheta_3(0|\tau) &= (-q;q^2)_\infty^2(q^2;q^2)_\infty \\
&= \frac{(-q;q^2)_\infty^2(q;q^2)_\infty^2}{(q;q^2)_\infty^2}(q^2;q^2)_\infty \\
&= \frac{(q^2;q^4)_\infty^2(q^2;q^2)_\infty^2}{(q;q)_\infty^2}(q^2;q^2)_\infty \\
&= \frac{(q^2;q^2)_\infty^2}{(q^4;q^4)_\infty^2}\frac{(q^2;q^2)_\infty^2}{(q;q)_\infty^2}(q^2;q^2)_\infty,
\end{aligned}$$

which implies that

$$\vartheta_3(0|\tau) = \frac{\eta^5(\tau)}{\eta^2(2\tau)\eta^2(\tau/2)}.$$

Using (6.29), we complete the proof of (6.30). $\qquad\square$

Remark 6.3. Identity (6.30) is used by Berggren, Borwein and Borwein [9, p. 689] to construct the number

$$\left(\frac{1}{10^{10}}\sum_{n=-\infty}^{\infty}e^{-n^2/10^{10}}\right)^2$$

which is different from π but coincides with the constant π to over 42 billion digits.

Remark 6.4. Let

$$SL_2(\mathbf{Z}) = \left\{ \begin{pmatrix} a & b \\ c & d \end{pmatrix} \middle| a, b, c, d \in \mathbf{Z}, ad - bc = 1 \right\}.$$

For $\begin{pmatrix} a & b \\ c & d \end{pmatrix} \in SL_2(\mathbf{Z})$ and $z \in \mathbf{C}$, we define "$\circ$" by

$$\begin{pmatrix} a & b \\ c & d \end{pmatrix} \circ z = \frac{az + b}{cz + d}.$$

Definition 6.14. Let k be a positive integer and let

$$\mathcal{H} = \{\tau \in \mathbf{C} | \operatorname{Im} \tau > 0\}.$$

A holomorphic function

$$f : \mathcal{H} \to \mathbf{C}$$

is a *modular form of weight k* on $SL_2(\mathbf{Z})$ if

$$f\left(\begin{pmatrix} a & b \\ c & d \end{pmatrix} \circ \tau \right) = f\left(\frac{a\tau + b}{c\tau + d} \right) = (c\tau + d)^k f(\tau) \qquad (6.31)$$

and

$$f(\tau) = \sum_{j=0}^{\infty} a_j e^{2\pi i j \tau}.$$

Theorem 6.9 shows that $E_{2j}(\tau) = L_{2j}(e^{2\pi i \tau})$, $j \geq 2$, is a modular form of weight $2j$ with respect to $SL_2(\mathbf{Z})$ while Theorem 6.10 shows that $\Delta(\tau)$ is a modular form of weight 12 with respect to $SL_2(\mathbf{Z})$. For more about modular forms with respect to $SL_2(\mathbf{Z})$, see the books by Serre [67, Chapter 7] and Koblitz [55, Chapter 3]. For applications of these modular forms to number theory, see for example the book by Apostol [5].

Exercises for Chapter 6

1. (a) Show that

$$\wp''(u|[\pi\tau,\pi]) = 6\wp^2(u|[\pi\tau,\pi]) - 30\mathcal{E}_4([\pi\tau,\pi]). \qquad (6.32)$$

(b) Deduce that

$$126\mathcal{E}_8([\pi\tau,\pi]) = 54\mathcal{E}_4^2([\pi\tau,\pi]).$$

(c) Use (6.17) to deduce that

$$L_8(x) = L_4^2(x).$$

(d) Use $L_8(x) = L_4^2(x)$ to conclude from (see (4.21))

$$-112L_4^2(x) - 320L_2(x)L_6(x) + 640x\frac{dL_6(x)}{dx} + 432L_8(x) = 0$$

that

$$x\frac{dL_6(x)}{dx} = \frac{L_2(x)L_6(x) - L_4^2(x)}{2}.$$

Remark: The above problem shows that we could avoid the use of the quintuple product identity in our proof of Theorem 4.4 using identities we derived in this chapter.

In [63, pp. 138–141], Ramanujan proved Theorem 4.4 by giving an elementary proof of (6.32) using trigonometric identities. His proof is closer to the proof given in the above problem.

2. Show that if $\tau \in \mathbf{C}$ with $\mathrm{Im}\,\tau > 0$, then

$$\vartheta_2(0\,|-1/\tau) = \sqrt{-i\tau}\vartheta_4(0|\tau). \qquad (6.33)$$

3. Let α, β, γ and δ be integers with $\alpha\delta - \gamma\beta = 1$. The Dedekind η-function satisfies the transformation formula

$$\eta\left(\frac{\alpha\tau + \beta}{\gamma\tau + \delta}\right) = \epsilon(\alpha,\beta,\gamma,\delta)\sqrt{\gamma\tau + \delta}\,\eta(\tau),$$

where $\epsilon(\alpha,\beta,\gamma,\delta)$ are certain 24th root of unity depending of α, β, γ and δ. Use this fact to show that if n, a, b, c and d are integers with $n \geq 2$, $n|c$ and $ad - bc = 1$, then

$$nE_2\left(n\frac{a\tau + b}{c\tau + d}\right) - E_2\left(\frac{a\tau + b}{c\tau + d}\right) = (c\tau + d)^2(nE_2(n\tau) - E_2(\tau)),$$

where

$$E_2(\tau) = L_2(e^{2\pi i\tau}).$$

4. Let a, b, c and d be integers satisfying $ad - bc = 1$.

    (a) Use (6.5) to show that

    $$\wp\big(u\,|\,[\pi\tau, \pi]\big) = \frac{1}{(c\tau + d)^2}\,\wp\left(\frac{u}{c\tau + d}\,\Big|\,\Big[\frac{a\tau + b}{c\tau + d}\pi, \pi\Big]\right). \qquad (6.34)$$

    (b) Use (6.34) and (6.15) to show that

    $$J\left(\frac{u}{c\tau + d}\,\Big|\,\frac{a\tau + b}{c\tau + d}\right) = (c\tau + d)^2 J(u\,|\,\tau) + C(\tau),$$

    where $C(\tau)$ is a constant independent of u.

    (c) Use (6.26) to show that

    $$C(\tau) = \frac{2c}{\pi i(c\tau + d)}.$$

7 An elliptic function of Jacobi

In this chapter, we introduce an elliptic function and derive identities associated with $E_4(\tau)$, $E_6(\tau)$, $\lambda(q)$ and $\vartheta_3(0|\tau)$. We will also give alternative proofs of the transformation formulae (6.30) and (6.33) satisfied by $\vartheta_j(0|\tau)$, $j = 2, 3$ and 4 and prove (6.10) for $\Delta(\tau)$ as a consequence. The proof of (6.10) using this approach does not require the knowledge of (4.15). Identities derived in this chapter will be used in the next chapter to connect theta functions with hypergeometric series.

7.1 An elliptic function of Jacobi

Let

$$B(u|\tau) = \left(\frac{\vartheta_4(u|\tau)}{\vartheta_1(u|\tau)} \right)^2.$$

Using Theorem 5.5, we deduce that $B(u|\tau)$ is an even elliptic function with periods π and $\pi\tau$.

Suppose

$$\frac{\vartheta_4(u|\tau)}{\vartheta_1(u|\tau)} = \frac{c_{-1}}{u} + c_0 + \cdots.$$

Then

$$c_{-1} = \frac{\vartheta_4(0|\tau)}{\vartheta_1'(0|\tau)}.$$

Using the identity (5.4), we conclude that

$$c_{-1} = \frac{1}{\vartheta_2(0|\tau)\vartheta_3(0|\tau)}.$$

This means that the coefficient of $1/u$ in the Laurent series expansion of $B(u/\zeta|\tau)$ is

$$c_{-1}\zeta = \frac{\vartheta_3(0|\tau)}{\vartheta_2(0|\tau)},$$

where

$$\zeta = \vartheta_3^2(0|\tau). \tag{7.1}$$

With these computations, we find that the coefficient of $1/u$ of the function

$$\frac{\vartheta_2(0|\tau)}{\vartheta_3(0|\tau)} \cdot \frac{\vartheta_4(u/\zeta|\tau)}{\vartheta_1(u/\zeta|\tau)}$$

is 1. This leads us to define the following function.

https://doi.org/10.1515/9783110541915-007

Definition 7.1. For $u, \tau \in \mathbf{C}$, with $\operatorname{Im} \tau > 0$, let

$$S(u|\tau) = \left(\frac{\vartheta_2(0|\tau)}{\vartheta_3(0|\tau)}\right)^2 \left(\frac{\vartheta_4(u/\zeta|\tau)}{\vartheta_1(u/\zeta|\tau)}\right)^2. \tag{7.2}$$

Note that the coefficient of $1/u^2$ in the expansion of $S(u|\tau)$ is 1 and that $S(u|\tau)$ is an even elliptic function with periods $\pi\zeta$ and $\pi\tau\zeta$, and a pole of order 2 at $u = 0$. The function $S(u|\tau)$ was first studied by Jacobi and it is usually written as

$$S(u|\tau) = \frac{1}{\{\operatorname{sn}(u, \mathbf{k})\}^2}.$$

The function $\operatorname{sn}(u, \mathbf{k})$ is one of Jacobi's elliptic functions (see [72, pp. 478–479] or [13, p. 57]) and $\mathbf{k} = \sqrt{\lambda(q)}$, with $q = e^{\pi i \tau}$.

7.2 A differential equation satisfied by $S(u|\tau)$

Theorem 7.2. *Let $S(u|\tau)$ be given by (7.2) and let*

$$\lambda = \lambda(e^{\pi i \tau}), \tag{7.3}$$

where $\lambda(q)$ is defined as in (2.5). Then

$$(S'(u))^2 = 4S(u|\tau)(S(u|\tau) - 1)(S(u|\tau) - \lambda). \tag{7.4}$$

Proof. We know that since $S'(u|\tau)$ is an odd elliptic function,

$$S'(-u|\tau) = -S'(u|\tau).$$

Hence, if ω is a period of $S(u|\tau)$, then

$$S'(\omega/2|\tau) = S'(\omega/2 - \omega|\tau)$$
$$= S'(-\omega/2|\tau)$$
$$= -S'(\omega/2|\tau).$$

Therefore,

$$S'(\omega/2|\tau) = 0$$

and so, the zeros of $S'(u|\tau)$ in the fundamental parallelogram $\mathcal{F}(\pi\zeta\tau, \pi\zeta)$ are

$$\pi\zeta/2, \quad \pi\tau\zeta/2$$

and

$$\pi(\tau + 1)\zeta/2,$$

where ζ is given by (7.1). With the above information and following the same argument as in the proof of Theorem 6.4, we conclude that

$$(S'(u|\tau))^2 = c(S(u|\tau) - S(\pi\zeta/2|\tau))(S(u|\tau) - S(\pi\tau\zeta/2|\tau)) \qquad (7.5)$$
$$\times (S(u|\tau) - S(\pi(\tau + 1)\zeta/2|\tau)).$$

Now, from Theorem 5.5, we deduce that

$$S(\pi\zeta/2|\tau) = \frac{\vartheta_2^2(0|\tau)\,\vartheta_4^2(\pi/2|\tau)}{\vartheta_3^2(0|\tau)\,\vartheta_1^2(\pi/2|\tau)}$$

$$= 1, \qquad (7.6)$$

$$S(\pi\tau\zeta/2|\tau) = \frac{\vartheta_2^2(0|\tau)\,\vartheta_4^2(\pi\tau/2|\tau)}{\vartheta_3^2(0|\tau)\,\vartheta_1^2(\pi\tau/2|\tau)}$$

$$= 0, \qquad (7.7)$$

and

$$S(\pi(\tau + 1)\zeta/2|\tau) = \frac{\vartheta_2^2(0|\tau)\,\vartheta_4^2(\pi(\tau + 1)/2|\tau)}{\vartheta_3^2(0|\tau)\,\vartheta_1^2(\pi(\tau + 1)/2|\tau)} = \frac{\vartheta_2^4(0|\tau)}{\vartheta_3^4(0|\tau)}$$

$$= \lambda. \qquad (7.8)$$

Identities (2.5) and (7.3) are used to derive the last equality of (7.8). Substituting (7.6), (7.7) and (7.8) in (7.5), we find that

$$(S'(u))^2 = cS(u|\tau)(S(u|\tau) - 1)(S(u|\tau) - \lambda).$$

By examining the expansion of $S(u|\tau)$, we conclude that $c = 4$ and this completes the proof of the theorem. □

We know that $S(u|\tau)$ is an even elliptic function with periods $\pi\zeta$ and $\pi\tau\zeta$ and it has only one pole at $u = 0$ of order 2. This implies that $S(u|\tau)$ and $\wp(u|[\pi\tau\zeta, \pi\zeta])$ differ by a constant which can be obtained by substituting $u = \pi\zeta/2$. The relation we obtained is

$$S(u|\tau) = \wp(u|[\pi\tau\zeta, \pi\zeta]) - \wp(\pi\zeta/2|[\pi\tau\zeta, \pi\zeta]). \qquad (7.9)$$

Substituting (7.9) in (7.6)–(7.8), we conclude that

$$\wp(\pi\zeta/2|[\pi\tau\zeta, \pi\zeta]) - \wp(\pi\tau\zeta/2|[\pi\tau\zeta, \pi\zeta]) = 1, \qquad (7.10)$$

$$\wp(\pi\zeta/2|[\pi\tau\zeta,\pi\zeta]) - \wp(\pi(1+\tau)\zeta/2|[\pi\tau\zeta,\pi\zeta]) = 1 - \lambda, \tag{7.11}$$

and

$$\wp(\pi\tau\zeta/2|[\pi\tau\zeta,\pi\zeta]) - \wp(\pi(\tau+1)\zeta/2|[\pi\tau\zeta,\pi\zeta]) = -\lambda. \tag{7.12}$$

From the definition of $\wp(u|[\pi\omega_1,\pi\omega_2])$ (see (6.1)), we observe that

$$\wp(u|[\pi\omega_1,\pi\omega_2]) = \frac{1}{\omega_2^2}\wp(u/\omega_2|[\pi\tau,\pi]), \tag{7.13}$$

where $\tau = \omega_1/\omega_2$. This implies that

$$\wp(u|[\pi\tau\zeta,\pi\zeta]) = \frac{1}{\zeta^2}\wp(u/\zeta|[\pi\tau,\pi])$$

and we may rewrite (7.10)–(7.12) as

$$\wp(\pi/2|[\pi\tau,\pi]) - \wp(\pi\tau/2|[\pi\tau,\pi]) = \zeta^2, \tag{7.14}$$

$$\wp(\pi/2|[\pi\tau,\pi]) - \wp(\pi(1+\tau)/2|[\pi\tau,\pi]) = \zeta^2(1-\lambda), \tag{7.15}$$

and

$$\wp(\pi\tau/2|[\pi\tau,\pi]) - \wp(\pi(1+\tau)/2|[\pi\tau,\pi]) = -\zeta^2\lambda. \tag{7.16}$$

Now, since

$$[\pi\tau,\pi] = [-\pi,\pi\tau],$$

we find that

$$\wp(u|[\pi\tau,\pi]) = \wp(u|[-\pi,\pi\tau]).$$

By (7.13), we deduce that

$$\wp(u|[\pi\tau,\pi]) = \frac{1}{\tau^2}\wp(u/\tau|[-\pi/\tau,\pi]).$$

This implies that

$$\wp(\pi\tau/2|[\pi\tau,\pi]) = \frac{1}{\tau^2}\wp(\pi/2|[-\pi/\tau,\pi]) \tag{7.17}$$

and

$$\wp(\pi/2|[\pi\tau,\pi]) = \frac{1}{\tau^2}\wp(\pi/(2\tau)|[-\pi/\tau,\pi])$$

$$= \frac{1}{\tau^2}\wp(-\pi/(2\tau)|[-\pi/\tau,\pi]), \tag{7.18}$$

where we have used the fact that $\wp$ is a even function of u in the last equality. Therefore, using (7.14), (7.17) and (7.18), we deduce that

$$\zeta^2(\tau) = \wp(\pi/2|[\pi\tau,\pi]) - \wp(\pi\tau/2|[\pi\tau,\pi])$$
$$= \frac{1}{\tau^2}\left(\wp(-\pi/(2\tau)|[-\pi/\tau,\pi]) - \wp\left(\frac{\pi}{2}|[\pi(-1/\tau),\pi]\right)\right)$$
$$= -\frac{1}{\tau^2}\zeta^2(-1/\tau).$$

This yields

$$\vartheta_3^4(0| - 1/\tau) = -\tau^2\vartheta_3^4(0|\tau). \tag{7.19}$$

Similarly, using (7.15) and (7.16), we deduce that

$$\vartheta_2^4(0| - 1/\tau) = -\tau^2\vartheta_4^4(0|\tau). \tag{7.20}$$

We have seen that the relation between $S(u|\tau)$ and Weierstrass' $\wp$ function leads to a proof of the transformation formula for $\vartheta_3(0|\tau)$ which we have encountered in Theorem 6.13. Theorem 6.13 was previously established using Theorem 6.12. The proof of Theorem 6.12, in turn, follows from Theorem 4.5 and Theorem 6.9.

We now give another proof of Theorem 6.12 using (7.19). This proof of Theorem 6.12 is independent of Theorem 4.5 and Theorem 6.9. The key to the proof of Theorem 6.12 is the identity (5.4) and Jacobi's triple product identity. Recall from (5.4) that

$$\vartheta_1'(0|\tau) = \vartheta_2(0|\tau)\vartheta_3(0|\tau)\vartheta_4(0|\tau).$$

By (7.19) and (7.20), we find that

$$(\vartheta_1'(0| - 1/\tau))^8 = \vartheta_2^8(0| - 1/\tau)\vartheta_3^8(0| - 1/\tau)\vartheta_4^8(0| - 1/\tau)$$
$$= \tau^{12}\vartheta_2^8(0|\tau)\vartheta_3^8(0|\tau)\vartheta_4^8(0|\tau)$$
$$= \tau^{12}(\vartheta_1'(0|\tau))^8. \tag{7.21}$$

From (3.5), we know that

$$\vartheta_1'(0|\tau) = 2q^{1/4}\prod_{k=1}^{\infty}(1 - q^{2k})^3$$

and hence (7.21) implies that

$$\Delta(-1/\tau) = \tau^{12}\Delta(\tau)$$

and this leads to the transformation formula (6.29) for $\eta(\tau)$, where $\eta(\tau)$ is given by (6.27).

7.3 The first few terms in the series expansion of $S(u|\tau)$

We know from the definition (7.2) of $S(u|\tau)$ that

$$S(u|\tau) = \frac{1}{u^2} + c_0(\tau) + c_2(\tau)u^2 + c_4(\tau)u^4 + \cdots,$$

where $c_{2j}(\tau), j \geq 0$ are independent of u. In this section, we will express $c_0(\tau), c_2(\tau)$ and $c_4(\tau)$ in terms of $\lambda = \lambda(e^{\pi i \tau})$ defined in (2.5). For simplicity, let $S(u) = S(u|\tau)$ and $c_j = c_j(\tau)$ for even nonnegative integer j.

By differentiating both sides of (7.4), we deduce that

$$S''(u) = 2((S(u) - 1)(S(u) - \lambda) + S(u)(S(u) - 1) + S(u)(S(u) - \lambda)). \tag{7.22}$$

Writing

$$S(u) = \frac{1}{u^2} + c_0 + c_2 u^2 + c_4 u^4 + c_6 u^6 + c_8 u^8 + \cdots,$$

we find that

$$S'(u) = -\frac{2}{u^3} + 2c_2 u + 4c_4 u^3 + \cdots$$

and

$$S''(u) = \frac{6}{u^4} + 2c_2 + 12c_4 u^2 + \cdots.$$

From (7.22), we deduce that

$$\frac{6}{u^4} + 2c_2 + 12c_4 u^2 + \cdots$$

$$= 2\left(\frac{1}{u^2} + c_0 - 1 + c_2 u^2 + c_4 u^4 + \cdots\right)\left(\frac{1}{u^2} + c_0 - \lambda + c_2 u^2 + c_4 u^4 + \cdots\right)$$

$$+ 2\left(\frac{1}{u^2} + c_0 + c_2 u^2 + c_4 u^4 + \cdots\right)\left(\frac{1}{u^2} + c_0 - \lambda + c_2 u^2 + c_4 u^4 + \cdots\right)$$

$$+ 2\left(\frac{1}{u^2} + c_0 + c_2 u^2 + c_4 u^4 + \cdots\right)\left(\frac{1}{u^2} + c_0 - 1 + c_2 u^2 + c_4 u^4 + \cdots\right). \tag{7.23}$$

Considering the coefficient of $1/u^2$ on both sides of (7.23), we conclude that

$$c_0 = \frac{\lambda + 1}{3}.$$

Similarly, comparing the constant term and the coefficients of u^2 on both sides of (7.23), we deduce that

$$c_2 = \frac{\lambda^2 - \lambda + 1}{15}.$$

and

$$c_4 = \frac{(\lambda - 2)(2\lambda - 1)(\lambda + 1)}{189}.$$

Hence, the following theorem holds.

Theorem 7.3. *The first few terms in the expansion of $S(u|\tau)$ about $u = 0$ is given by*

$$S(u|\tau) = \frac{1}{u^2} + \frac{\lambda + 1}{3} + \frac{1 - \lambda + \lambda^2}{15}u^2 + \frac{(\lambda - 2)(2\lambda - 1)(\lambda + 1)}{189}u^4 + \cdots. \qquad (7.24)$$

7.4 Representations of Eisenstein series $E_4(\tau)$ and $E_6(\tau)$ in terms of $\vartheta_3(0|\tau)$ and λ

We have seen from (7.2) that $S(u|\tau)$ is an even elliptic function with periods $\pi\zeta$ and $\pi\tau\zeta$, where $\zeta = \vartheta_3^2(0|\tau)$ with pole of order 2 at $u = 0$.

Now, the function $J(u/\zeta|\tau)$, where $J(u|\tau)$ is defined in (6.12), is also an elliptic function with periods $\pi\zeta$ and $\pi\tau\zeta$ and it has a pole of order 2 at $u = 0$. By Theorem 5.3, there exist a, b which are dependent on τ but independent of u such that

$$S(u|\tau) = aJ(u/\zeta|\tau) + b. \qquad (7.25)$$

Next, using (6.14), we find that

$$J(u/\zeta|\tau) = -\frac{\zeta^2}{u^2} + \sum_{k=1}^{\infty} \frac{(-1)^k}{2k} \frac{B_{2k}L_{2k}(e^{2\pi i\tau})2^{2k}}{(2k-2)!\zeta^{2k-2}}u^{2k-2}$$

$$= -\frac{\zeta^2}{u^2} + \sum_{k=1}^{\infty} \frac{(-1)^k}{2k} \frac{B_{2k}E_{2k}(\tau)2^{2k}}{(2k-2)!\zeta^{2k-2}}u^{2k-2}, \qquad (7.26)$$

where we have replaced L_{2k} with E_{2k} using (6.18). Using (7.24), we know that

$$S(u|\tau) = \frac{1}{u^2} + \frac{\lambda + 1}{3} + \frac{1 - \lambda + \lambda^2}{15}u^2 + \frac{(\lambda - 2)(2\lambda - 1)(\lambda + 1)}{189}u^4 + \cdots. \qquad (7.27)$$

Using (7.26) and (7.27) in (7.25), we deduce that

$$a = -\frac{1}{\zeta^2},$$

$$b = \frac{\lambda + 1}{3} - \frac{E_2(\tau)}{3\zeta^2},$$

$$-2^4 \cdot 3 \cdot \frac{B_4}{4!} \frac{L_4}{\vartheta_3^8(0|\tau)} = \frac{1}{15}(1 - \lambda + \lambda^2),$$

and

$$2^6 \cdot 5 \cdot \frac{B_6}{6!} \frac{L_6}{g_3^{12}} = \frac{1}{189}(\lambda - 2)(2\lambda - 1)(\lambda + 1).$$

Simplifying, we conclude that

$$E_4(\tau) = \zeta^4(1 - \lambda + \lambda^2) \tag{7.28}$$

and

$$E_6(\tau) = \zeta^6(1 + \lambda)(2 - \lambda)(1 - 2\lambda)/2. \tag{7.29}$$

Remark 7.1. Identities (7.28) and (7.29) can also be obtained using the relation between $S(u|\tau)$ and $\wp(u/\zeta|[\pi\tau, \pi])$, namely (7.9) and the expansion of Weierstrass' $\wp$ function given by (6.16).

Exercises for Chapter 7

1. Complete the proof of (7.20).
2. Use (2.23), (7.19) and (7.20) to show that, for $\operatorname{Im} \tau > 0$,

$$\lambda(e^{-i\pi/\tau}) = 1 - \lambda(e^{i\pi\tau}). \tag{7.30}$$

3. Complete the following problems which lead to a proof of (4.15) that is independent of the system of differential equations in Theorem 4.4.

   (a) We have seen that $\wp(u|[\pi\tau, \pi])$ satisfies the differential equation

   $$\left(\wp'(u|[\pi\tau, \pi])\right)^2 = 4\wp^3(u|[\pi\tau, \pi]) - 60\mathcal{E}_4([\pi\tau, \pi])\wp(u|[\pi\tau, \pi]) - 140\mathcal{E}_6([\pi\tau, \pi]).$$

   Use (6.17) and (6.18) to rewrite (7.31) as

   $$\left(\wp'(u|[\pi\tau, \pi])\right)^2 = 4\wp^3(u|[\pi\tau, \pi]) - \frac{4}{3}E_4(\tau)\wp(u|[\pi\tau, \pi]) - \frac{8}{27}E_6(\tau). \tag{7.31}$$

   (b) Let μ, v and ω be the roots of the polynomial $p(x) = x^3 + bx + c$. The discriminant of the polynomial $p(x)$ is defined as

   $$(\mu - v)^2(\mu - \omega)^2(v - \omega)^2.$$

   Show that

   $$(\mu - v)^2(\mu - \omega)^2(v - \omega)^2 = -4b^3 - 27c^2.$$

   (c) Show that the right-hand side of (7.31) can be expressed as a product, namely,

   $$\begin{aligned}\left(\wp'(u|[\pi\tau, \pi])\right)^2 = {}& 4(\wp(u|[\pi\tau, \pi]) - \wp(\pi/2|[\pi\tau, \pi])) \\ & \times (\wp(u|[\pi\tau, \pi]) - \wp(\pi\tau/2|[\pi\tau, \pi])) \\ & \times (\wp(u|[\pi\tau, \pi]) - \wp(\pi(\tau + 1)/2|[\pi\tau, \pi])).\end{aligned} \tag{7.32}$$

   (d) Use (b), (c), (7.31) and (7.32) to show that

   $$\begin{aligned}& \left(\wp(\pi/2|[\pi\tau, \pi]) - \wp(\pi\tau/2|[\pi\tau, \pi])\right)^2 \\ & \quad \times \left(\wp(\pi/2|[\pi\tau, \pi]) - \wp(\pi(\tau + 1)/2|[\pi\tau, \pi])\right)^2 \\ & \quad \times \left(\wp(\pi\tau/2|[\pi\tau, \pi]) - \wp(\pi(\tau + 1)/2|[\pi\tau, \pi])\right)^2 \\ & = \frac{4}{27}\left(E_4(\tau)^3 - E_6(\tau)^2\right)\end{aligned}$$

   and deduce from (7.14), (7.15), (7.16) and (1.11) that

   $$E_4^3(\tau) - E_6^2(\tau) = 1728\Delta(\tau).$$

Remark 7.2. We have seen that $\Delta(\tau)$ can be written as the discriminant of a certain cubic polynomial. This is why it is sometimes called the modular discriminant.

4. In this problem, we show that it is enough to determine the expansion of $S(u|\tau)$ up to u^2 in order to derive expressions of $E_4(\tau)$ and $E_6(\tau)$ in terms of ζ and λ.

    (a) Show that

    $$\Delta(\tau) = \frac{1}{256}\zeta^{12}\lambda^2(1-\lambda)^2 \tag{7.33}$$

    using

    $$\vartheta_1'(0|\tau) = \vartheta_2(0|\tau)\vartheta_3(0|\tau)\vartheta_4(0|\tau).$$

    (b) Deduce that if we have the expression

    $$E_4(\tau) = \zeta^4(1-\lambda+\lambda^2),$$

    then we may derive the identity

    $$E_6(\tau) = \zeta^6(1+\lambda)(2-\lambda)(1-2\lambda)/2$$

    from (7.33) and the relation

    $$\Delta(\tau) = \frac{1}{1728}(E_4^3(\tau) - E_6^2(\tau)).$$

5. Use (2.32), (2.2) and (2.26) to show that

    $$\lambda(q) = 1 - \left(\frac{1-\sqrt{\lambda(q^2)}}{1+\sqrt{\lambda(q^2)}}\right)^2,$$

    $$\vartheta_3^2(0|\tau/2) = \vartheta_3^2(0|\tau) + \vartheta_2^2(0|\tau),$$

    $$\vartheta_3^2(0|\tau/2) = \vartheta_3^2(0|\tau)\left(1 + \frac{\vartheta_2^2(0|\tau)}{\vartheta_3^2(0|\tau)}\right),$$

    $$2\vartheta_3^2(0|2\tau) = \vartheta_3^2(0|\tau) + \vartheta_4^2(0|\tau),$$

    and

    $$\vartheta_3^2(0|2\tau) = \frac{\vartheta_3^2(0|\tau)}{2}\left(1 + \frac{\vartheta_4^2(0|\tau)}{\vartheta_3^2(0|\tau)}\right).$$

6. Use (7.28) and the identities in Problem 5 to deduce that

$$E_4(\tau/2) = \zeta^4(1 + 14\lambda + \lambda^2)$$

and

$$E_4(2\tau) = \zeta^4(1 - \lambda + \lambda^2/16).$$

7. Conclude that

$$\vartheta_3^8(0|\tau) = \frac{E_4(\tau/2)}{15} - \frac{2}{15}E_4(\tau) + \frac{16}{15}E_4(2\tau).$$

8. In Chapter 5, one identity that plays an important role in our proof of the identity (see (5.4))

$$\vartheta_1'(0|\tau) = \vartheta_2(0|\tau)\vartheta_3(0|\tau)\vartheta_4(0|\tau)$$

is the identity

$$\vartheta_2^2(0|\tau) = 2\vartheta_2(0|2\tau)\vartheta_3(0|2\tau), \tag{7.34}$$

which we derived by elementary manipulations of identities satisfied by Jacobi's theta functions. Use the transformation formulas (7.19) and (7.20) to show that (7.34) follows from (see Theorem 2.1)

$$\vartheta_4^2(0|2\tau) = \vartheta_3(0|\tau)\vartheta_4(0|\tau).$$

8 Hypergeometric series and Ramanujan's series $1/\pi$

In this chapter, we give a proof of Lagrange's four-square theorem using Ramanujan's differential equations and the parametrization of $L_4(x)$ and $L_6(x)$ in terms of ζ and λ. Using the same set of identities, we derive a second order differential equation satisfied by ζ and λ, allowing us to express ζ as a hypergeometric series in λ.

8.1 Ramanujan's differential equations and Lagrange's four-square theorem

Let $x = e^{2\pi i \tau} = q^2$. From (7.28) and (7.29), the functions $L_4(x) = E_4(\tau)$ and $L_6(x) = E_6(\tau)$ can be expressed in the form

$$L_4(x) = \zeta^4(1 - \lambda + \lambda^2) \tag{8.1}$$

and

$$L_6(x) = \zeta^6(1 + \lambda)\left(1 - \frac{\lambda}{2}\right)(1 - 2\lambda), \tag{8.2}$$

where

$$\lambda = \frac{\vartheta_2^4(0|\tau)}{\vartheta_3^4(0|\tau)}$$

and

$$\zeta = \vartheta_3^2(0|\tau).$$

Using (8.1) and (8.2), we deduce that

$$x\frac{dL_4(x)}{dx} = q\frac{dL_4(x)}{2dq}$$

$$= 2\zeta^3(1 - \lambda + \lambda^2)q\frac{d\zeta}{dq} + \frac{\zeta^4}{2}(-1 + 2\lambda)q\frac{d\lambda}{dq} \tag{8.3}$$

and

$$x\frac{dL_6(x)}{dx} = q\frac{dL_6(x)}{2dq}$$

$$= 3\zeta^5(1 + \lambda)(1 - \lambda/2)(1 - 2\lambda)q\frac{dz}{dq} + \frac{3}{2}\zeta^6(-1/2 - \lambda + \lambda^2)q\frac{d\lambda}{dq}. \tag{8.4}$$

https://doi.org/10.1515/9783110541915-008

From (8.1), (8.2), (8.3) and (8.4), we deduce that

$$3L_6(x)x\frac{dL_4(x)}{dx} - 2L_4(x)x\frac{dL_6(x)}{dx} = \frac{27}{4}\zeta^{10}\lambda(1-\lambda)q\frac{d\lambda}{dq}. \tag{8.5}$$

Next, from Theorem 4.4, we know that

$$x\frac{dL_2(x)}{dx} = \frac{L_2^2(x) - L_4(x)}{12}, \tag{8.6}$$

$$x\frac{dL_4(x)}{dx} = \frac{L_2(x)L_4(x) - L_6(x)}{3}, \tag{8.7}$$

and

$$x\frac{dL_6(x)}{dx} = \frac{L_2(x)L_6(x) - L_4^2(x)}{2}. \tag{8.8}$$

From (8.6) to (8.8), we deduce that

$$3L_6(x)x\frac{dL_4(x)}{dx} - 2L_4(x)x\frac{dL_6(x)}{dx} = L_4^3(x) - L_6^2(x). \tag{8.9}$$

By (8.1) and (8.2), we find that the right-hand side of (8.9) is given by

$$L_4^3(x) - L_6^2(x) = \frac{27}{4}\zeta^{12}(\lambda(1-\lambda))^2. \tag{8.10}$$

Simplifying (8.9) using (8.5) and (8.10), we deduce that

$$q\frac{d\lambda}{dq} = \zeta^2\lambda(1-\lambda), \tag{8.11}$$

or

$$2x\frac{d\lambda}{dx} = \zeta^2\lambda(1-\lambda), \tag{8.12}$$

where $x = q^2$.

Identity (8.11) is equivalent to an identity that leads to a proof of Lagrange's four-square theorem.

We now state Lagrange's theorem and give a proof using (8.11).

Theorem 8.1. *Every positive integer can be written as a sum of four squares.*

Proof. Using (3.3) and (3.4), we find that

$$\lambda = \frac{\vartheta_2^4(0|\tau)}{\vartheta_3^4(0|\tau)} = 16q\prod_{k=1}^{\infty}\frac{(1+q^{2k})^8}{(1+q^{2k-1})^8}, \tag{8.13}$$

where $q = e^{\pi i \tau}$. Note that the product on the right-hand side of (8.13) can be written as

$$\prod_{k=1}^{\infty} \frac{(1 + q^{2k})^8}{(1 + q^{2k-1})^8} = \prod_{k=1}^{\infty} \left(\frac{1}{(1 - q^{4k-2})(1 + q^{2k-1})} \right)^8. \tag{8.14}$$

Substituting (8.14) into (8.13) and logarithmically differentiating both sides of the resulting identity, we deduce that

$$\frac{1}{\lambda} q \frac{d\lambda}{dq} = 1 + 8 \sum_{k=1}^{\infty} \left(\frac{(4k-2)q^{4k-2}}{1 - q^{4k-2}} - \frac{(2k-1)q^{2k-1}}{1 + q^{2k-1}} \right). \tag{8.15}$$

By (8.11) and (2.23), we may rewrite the left-hand side of (8.15) as

$$\begin{aligned}
\frac{1}{\lambda} q \frac{d\lambda}{dq} &= \zeta^2(1 - \lambda) \\
&= \vartheta_3^4(0|\tau) \left(1 - \frac{\vartheta_2^4(0|\tau)}{\vartheta_3^4(0|\tau)} \right) \\
&= \vartheta_4^4(0|\tau).
\end{aligned}$$

Therefore,

$$\left(\sum_{k=-\infty}^{\infty} (-1)^k q^{k^2} \right)^4 = 1 + 8 \sum_{k=1}^{\infty} \left(\frac{(4k-2)q^{4k-2}}{1 - q^{4k-2}} - \frac{(2k-1)q^{2k-1}}{1 + q^{2k-1}} \right).$$

Replacing $-q$ by q, we conclude that

$$\begin{aligned}
\left(\sum_{k=-\infty}^{\infty} q^{k^2} \right)^4 &= 1 + 8 \sum_{k=1}^{\infty} \left(\frac{(4k-2)q^{4k-2}}{1 - q^{4k-2}} + \frac{(2k-1)q^{2k-1}}{1 - q^{2k-1}} \right) \\
&= 1 + 8 \sum_{n=1}^{\infty} \left(\sum_{\substack{d|n \\ d \equiv 2 \, (\mathrm{mod}\ 4)}} d + \sum_{\substack{d|n \\ d \equiv 1 \, (\mathrm{mod}\ 2)}} d \right) q^n \\
&= 1 + 8 \sum_{n=1}^{\infty} \left(\sum_{\substack{d|n \\ 4 \nmid d}} d \right) q^n. \tag{8.16}
\end{aligned}$$

If we write

$$\left(\sum_{k=-\infty}^{\infty} q^{k^2} \right)^4 = \sum_{n=0}^{\infty} r_4(n) q^n,$$

then from (8.16), we conclude that

$$r_4(n) = 8 \sum_{\substack{d|n \\ 4\nmid d}} d. \tag{8.17}$$

Since every positive integer n is divisible by 1, (8.17) implies that $r_4(n) \geq 8$ and hence, every positive integer n is a sum of four squares. $\qquad\square$

Remark 8.1. Our proof of (8.16) that leads to Lagrange's four-square theorem appears to be new. For a different proof of (8.16) and many elegant identities associated with the representations of integers as sums of squares, we encourage the reader to consult the book by Grosswald [45].

In 1994, Kac and Wakimoto [53] conjectured formula for the number of representations of n as sums of v squares, where $v = 4s^2$ and $4s(s+1)$. Their conjectures were proved by Milne [60] using explicit identities that involved determinants of matrices with entries in terms of $E_{2j}(\tau)$. In 2000, Zagier [75] proved Milne's formulas using theory of modular forms. For more details, see the survey paper by Chan and Krattenthaler [32].

8.2 Hypergeometric series and a differential equation satisfied by $\vartheta_3(0|\tau)$ and λ

In this section, we will derive a differential equation satisfied by ζ and λ and connect our study of these functions with hypergeometric series.

The identity (8.12) allows us to rewrite differential equations (8.6)–(8.8) as

$$\frac{\zeta^2\lambda(1-\lambda)}{2}\frac{dL_2(x)}{d\lambda} = \frac{L_2^2(x) - L_4(x)}{12}, \tag{8.18}$$

$$\frac{\zeta^2\lambda(1-\lambda)}{2}\frac{dL_4(x)}{d\lambda} = \frac{L_2(x)L_4(x) - L_6(x)}{3}, \tag{8.19}$$

and

$$\frac{\zeta^2\lambda(1-\lambda)}{2}\frac{dL_6(x)}{d\lambda} = \frac{L_2(x)L_6(x) - L_4^2(x)}{2}.$$

From (8.1), we deduce that

$$\frac{dL_4(x)}{d\lambda} = 4\zeta^3(1 - \lambda + \lambda^2)\frac{d\zeta}{d\lambda} + \zeta^4(2\lambda - 1). \tag{8.20}$$

Using (8.19), (8.20) and (8.2), we can solve for L_2 and derive the "missing" parametrization of L_2 in terms of ζ and λ, namely,

$$L_2(x) = 6\lambda(1 - \lambda)\zeta\frac{d\zeta}{d\lambda} + \zeta^2(1 - 2\lambda). \tag{8.21}$$

Next, using (8.21) and (8.18) to eliminate L_2, we complete the proof of the following theorem.

Theorem 8.2. *Let* $\operatorname{Im}\tau > 0$,

$$\zeta = \vartheta_3^2(0|\tau),$$

and

$$\lambda = \frac{\vartheta_2^4(0|\tau)}{\vartheta_3^4(0|\tau)}.$$

Then

$$\lambda(1-\lambda)\frac{d^2\zeta}{d\lambda^2} + \frac{d\zeta}{d\lambda}(1-2\lambda) - \frac{\zeta}{4} = 0. \tag{8.22}$$

Let

$$D_t = t\frac{d}{dt}.$$

Then

$$D_t f = t\frac{df(t)}{dt} \tag{8.23}$$

and

$$D_t^2 f - D_t f = t^2\frac{d^2f(t)}{dt^2}. \tag{8.24}$$

Using (8.23) and (8.24), we may rewrite (8.22) as

$$(1-\lambda)(D_\lambda^2\zeta - D_\lambda\zeta) + (1-2\lambda)D_\lambda\zeta - \frac{\lambda\zeta}{4} = 0,$$

which upon simplification yields

$$D_\lambda^2\zeta = \lambda\left(D_\lambda + \frac{1}{2}\right)^2\zeta. \tag{8.25}$$

By writing

$$\zeta = \sum_{n=0}^{\infty} a_n\lambda^n, \tag{8.26}$$

(8.25) implies that

$$n^2 a_n = \left(\frac{1}{2} + n - 1\right)^2 a_{n-1} \tag{8.27}$$

and this implies that

$$\zeta = \vartheta_3^2(0|\tau) = a_0 \cdot {}_2F_1\left(\frac{1}{2}, \frac{1}{2}; 1; \lambda\right), \tag{8.28}$$

where *the hypergeometric series* ${}_pF_q(a_1, \ldots, a_p; b_1, \ldots, b_q; u)$ is defined by

$$_pF_q(a_1, \ldots, a_p; b_1, \ldots, b_q; u) = \sum_{k=0}^{\infty} \frac{(a_1)_k \cdots (a_p)_k}{(b_1)_k \cdots (b_q)_k} \frac{u^k}{k!},$$

with

$$(\ell)_0 = 1 \quad \text{and for } k \ge 1, \quad (\ell)_k = \ell(\ell + 1) \cdots (\ell + k - 1).$$

By comparing the series expansion in q on both sides using the series expansions of ζ and λ, we conclude that $a_0 = 1$.

Theorem 8.3. *For* $\operatorname{Im} \tau > 0$ *with* $|\lambda| = |\lambda(e^{\pi i \tau})| < 1$, *we have*

$$\vartheta_3^2(0|\tau) = {}_2F_1\left(\frac{1}{2}, \frac{1}{2}; 1; \lambda\right). \tag{8.29}$$

Remark 8.2. We derive Theorem 8.3 by assuming (8.26). To avoid this assumption, we recall that the general solutions of the differential equation [41, p. 246]

$$(1 - u)(D_u^2 y - D_u y) + (1 - 2u)D_u y - \frac{u}{4} y = 0$$

is given by

$$y = A\, {}_2F_1\left(\frac{1}{2}, \frac{1}{2}; 1; u\right) + B\, {}_2F_1\left(\frac{1}{2}, \frac{1}{2}; 1; 1 - u\right), \tag{8.30}$$

where A, B are certain complex numbers. The construction of ${}_2F_1(\frac{1}{2}, \frac{1}{2}; 1; u)$ from (8.30) follows exactly the steps we have taken in the derivation of (8.28) from (8.27). Using (8.30), we conclude that

$$\zeta = A\, {}_2F_1\left(\frac{1}{2}, \frac{1}{2}; 1; \lambda\right) + B\, {}_2F_1\left(\frac{1}{2}, \frac{1}{2}; 1; 1 - \lambda\right)$$

for some constants A and B. By comparing the series expansion in q of both sides using the definitions of ζ and λ, we conclude that $A = 1$ and $B = 0$. This completes the proof of Theorem 8.3.

Theorem 8.3 expresses ζ as a power series in λ. Next, we express ζ as a power series in

$$X = 4\lambda(1 - \lambda). \tag{8.31}$$

Observe that from (8.31)

$$1 - 2\lambda = \sqrt{1 - X}$$

and we can rewrite (8.25) as

$$X\frac{d^2\zeta}{d\lambda^2} + 4\sqrt{1 - X}\frac{d\zeta}{d\lambda} - \zeta = 0. \tag{8.32}$$

Using

$$\frac{dX}{d\lambda} = 4\sqrt{1 - X},$$

we rewrite (8.32) as

$$X(1 - X)\frac{d^2\zeta}{dX^2} + \left(1 - \frac{3}{2}X\right)\frac{d\zeta}{dX} - \frac{\zeta}{16} = 0. \tag{8.33}$$

Expressing (8.33) in terms of D_X (see (8.23)), we find that

$$D_X^2\zeta = X\left(D_X + \frac{1}{4}\right)^2\zeta. \tag{8.34}$$

Identity (8.34) is similar to (8.25) and following the method of deriving (8.29) from (8.25), we deduce from (8.34) that

$$\zeta = {}_2F_1\left(\frac{1}{4}, \frac{1}{4}; 1; X\right). \tag{8.35}$$

Remark 8.3. From identities (8.29) and (8.35), we obtain the following quadratic transformation formula for ${}_2F_1(\frac{1}{2}, \frac{1}{2}; 1; t)$:

$$
{}_2F_1\left(\frac{1}{2}, \frac{1}{2}; 1; t\right) = {}_2F_1\left(\frac{1}{4}, \frac{1}{4}; 1; 4t(1 - t)\right). \tag{8.36}
$$

Identity (8.36) is a special case of

$$
{}_2F_1\left(2a, 2b; a + b + \frac{1}{2}; t\right) = {}_2F_1\left(a, b; a + b + \frac{1}{2}; 4t(1 - t)\right), \tag{8.37}
$$

which can be proved without the considerations of theta functions. For more details of the proof of (8.37), see [3, p. 125].

Now, if we let $Z = \zeta^2$, then

$$\zeta = Z^{1/2}.$$

Since

$$D_X \zeta = \frac{1}{2Z^{1/2}} D_X Z$$

and

$$D_X^2 \zeta = -\frac{1}{4Z^{3/2}} (D_X Z)^2 + \frac{1}{2Z^{1/2}} D_X^2 Z,$$

we can rewrite (8.34) as

$$(1 - X)(-(D_X Z)^2 + 2Z D_X^2 Z) = Z X D_X Z + \frac{X Z^2}{4}.$$

Applying D_X to both sides of the above differential equation, we find that

$$-X(-(D_X Z)^2 + 2Z D_X^2 Z) + 2(1 - X)Z D_X^3 Z$$
$$= \frac{3}{2} Z X D_X Z + X (D_X Z)^2 + Z X D_X^2 Z + \frac{X Z^2}{4},$$

which simplifies to

$$D_X^3 Z = X \left(D_X + \frac{1}{2} \right)^3 Z. \tag{8.38}$$

Once again, (8.38) is similar to (8.25) and (8.34) and we deduce, in exactly the same manner as obtaining (8.29) from (8.25), that (8.38) implies that

$$Z = {}_3F_2\left(\frac{1}{2}, \frac{1}{2}, \frac{1}{2}; 1, 1; X \right). \tag{8.39}$$

Remark 8.4. Clausen's identity [3, p. 116] states that

$$\left\{ {}_2F_1\left(a, b; a + b + \frac{1}{2}; u \right) \right\}^2 = {}_3F_2\left(2a, 2b, a + b; 2a + 2b, a + b + \frac{1}{2}; u \right). \tag{8.40}$$

When $a = b = 1/4$, (8.40) reduces to

$${}_2F_1^2\left(\frac{1}{4}, \frac{1}{4}; 1; u \right) = {}_3F_2\left(\frac{1}{2}, \frac{1}{2}, \frac{1}{2}; 1, 1; u \right). \tag{8.41}$$

Identity (8.41) follows from the combination of (8.35) and (8.39). The identity (8.41) can also be written, using (8.29), (8.35) and (8.31), as

$${}_2F_1^2\left(\frac{1}{2}, \frac{1}{2}; 1; x \right) = {}_3F_2\left(\frac{1}{2}, \frac{1}{2}, \frac{1}{2}; 1, 1; 4x(1 - x) \right). \tag{8.42}$$

Remark 8.5. In [17] (see also [3, p. 400]), Brafman discovered the identity

$$\sum_{j=0}^{\infty} \frac{(\gamma)_j(\alpha + \beta - \gamma + 1)_j P_j^{(\alpha,\beta)}(x)r^j}{(\alpha + 1)_j(\beta + 1)_j}$$

$$= {}_2F_1(\gamma, \alpha + \beta - \gamma + 1; \alpha + 1; (1 - r - R)/2)$$
$$\times {}_2F_1(\gamma, \alpha + \beta - \gamma + 1; \beta + 1; (1 + r - R)/2), \tag{8.43}$$

where

$$R = \left(1 - 2xr + r^2\right)^{1/2}$$

and

$$P_n^{(\alpha,\beta)}(x) = \frac{(\alpha + 1)_n}{(1)_n} {}_2F_1(-n, n + \alpha + \beta + 1, \alpha + 1; (1 - x)/2). \tag{8.44}$$

Brafman's identity (8.43) is a generalization of (8.42). For more details, see [70, p. 489]. The function $P_n^{(\alpha,\beta)}(x)$ defined in (8.44) is known as (see [3, p. 99]) *the Jacobi polynomial of degree n.*

Recently, motivated by a collection of series for $1/\pi$ discovered by Sun (see [35]) and several new analogues of (8.42) found using modular forms in [34], Wan and Zudilin [70] (see also [27]) derived an elegant generalization of Brafman's identity. We state their result without proof as follows.

Theorem 8.4. *Let $a, b, c \in \mathbf{Q}$. Let $u_{-1} = 0, u_0 = 1$,*

$$(n + 1)^2 u_{n+1} = (an^2 + an + b)u_n - cn^2 u_{n-1}.$$

Then in a neighborhood of $(X, Y) = (0, 0)$,

$$G(X)G(Y) = H(X, Y),$$

where

$$G(Z) = \sum_{k=0}^{\infty} u_k Z^k,$$

$$H(X, Y) = \frac{1}{W} \sum_{k=0}^{\infty} u_k \sum_{j=0}^{k} \binom{k}{j}^2 \left(\frac{X(1 - aY + cY^2)}{(1 - cXY)^2}\right)^j \left(\frac{Y(1 - aX + cX^2)}{(1 - cXY)^2}\right)^{k-j}.$$

A generalization of the identity found by Wan and Zudilin was later discovered by Chan and Tanigawa [33].

When $a = 1$, $b = 1/4$, $c = 0$ and $X = Y$, Theorem 8.4 yields (8.41). When $a = 7$, $b = 2$, $c = -8$, $X = Y$, one obtains [34]

$$\left(\sum_{k=0}^{\infty} \sum_{j=0}^{k} \binom{k}{j}^3 z^k \right)^2 = \frac{1}{1 + 8x^2} \sum_{k=0}^{\infty} \binom{2k}{k} \sum_{j=0}^{k} \binom{k}{j}^3 \left(\frac{z(1+z)(1-8z)}{(1+8z^2)^2} \right)^k,$$

which is an identity first derived using the theory of modular forms.

8.3 Hypergeometric series and π

Let r be a positive real number. We first recall by (8.29) that

$$\Theta_3^2(e^{-\pi/r}) = {}_2F_1\left(\frac{1}{2}, \frac{1}{2}; 1; \lambda(e^{-\pi/r}) \right) = {}_2F_1\left(\frac{1}{2}, \frac{1}{2}; 1; 1 - \lambda(e^{-\pi r}) \right). \tag{8.45}$$

On the other hand, by (6.30),

$$\Theta_3^2(e^{-\pi/r}) = r\Theta_3^2(e^{-\pi r}) = r\,{}_2F_1\left(\frac{1}{2}, \frac{1}{2}; 1; \lambda(e^{-\pi r}) \right). \tag{8.46}$$

Combining (8.45) and (8.46), we deduce the following theorem.

Theorem 8.5. *Let*

$$F(t) = {}_2F_1\left(\frac{1}{2}, \frac{1}{2}; 1; t \right).$$

For $r \in \mathbf{R}^+$, let

$$\lambda_r = \lambda(e^{-\pi r}). \tag{8.47}$$

Then

$$r = \frac{F(1 - \lambda_r)}{F(\lambda_r)}. \tag{8.48}$$

Differentiating (8.48) with respect to λ_r, we deduce that

$$\frac{1}{\pi} = \left(\frac{F'(1-\lambda_r)}{F(\lambda_r)} + \frac{F'(\lambda_r)F(1-\lambda_r)}{F^2(\lambda_r)} \right) q \frac{d\lambda_r}{dq}, \tag{8.49}$$

where

$$F'(u) = \frac{dF(u)}{du} \quad \text{and} \quad q = e^{-\pi r}.$$

By (8.11) and (8.29), we know that

$$q\frac{d\lambda_r}{dq} = F^2(\lambda_r)\lambda_r(1 - \lambda_r). \tag{8.50}$$

Substituting (8.50) into (8.49), we conclude the proof of the following theorem.

Theorem 8.6. *Let $r > 0$ and $\lambda_r = \lambda(e^{-\pi r})$. Then*

$$\frac{1}{\pi} = \lambda_r(1 - \lambda_r)(F'(1 - \lambda_r)F(\lambda_r) + F'(\lambda_r)F(1 - \lambda_r)). \tag{8.51}$$

Recall from (2.10) that

$$\Theta_3^2(q) + \Theta_4^2(q) = 2\Theta_3^2(q^2). \tag{8.52}$$

Using (8.29) and (2.23), we express (8.52) as

$$2F(\beta) = (1 + \sqrt{1 - \lambda})F(\lambda), \tag{8.53}$$

where $\beta = \lambda(q^2)$. Differentiating (8.53) with respect to λ, we find that

$$-\frac{1}{2\sqrt{1 - \lambda}}F(\lambda) + (1 + \sqrt{1 - \lambda})F'(\lambda) = 2F'(\beta)\frac{d\beta}{d\lambda}. \tag{8.54}$$

But by (8.11) with q replaced by q^2, we find that

$$q\frac{d\beta}{dq} = 2F^2(\beta)\beta(1 - \beta),$$

which implies that

$$\frac{d\beta}{d\lambda} = 2\frac{\beta(1 - \beta)}{\lambda(1 - \lambda)}\frac{F^2(\beta)}{F^2(\lambda)}. \tag{8.55}$$

Substituting (8.53) and (8.55) into (8.54), we complete the proof of the following theorem.

Theorem 8.7. *Let $\lambda = \lambda(q)$ and $\beta = \lambda(q^2)$. Then*

$$-\frac{\lambda(1 - \lambda)}{\sqrt{1 - \lambda}}F(\lambda) + 2\lambda(1 - \lambda)F'(\lambda)(1 + \sqrt{1 - \lambda}) = 2\beta(1 - \beta)F'(\beta)(1 + \sqrt{1 - \lambda})^2. \tag{8.56}$$

8.4 Ramanujan's series for $1/\pi$

In this section, we will derive a series for $1/\pi$, which is originally due to Ramanujan. We need to first record several identities.

Let $q = e^{-\pi/\sqrt{2}}$. Then

$$\beta(e^{-\pi/\sqrt{2}}) = \lambda(e^{-2\pi/\sqrt{2}}) = \lambda(e^{-\pi\sqrt{2}}) = 1 - \lambda(e^{-\pi/\sqrt{2}}), \tag{8.57}$$

where the last equality follows from (7.30). Identity (8.57) implies that

$$\lambda_{\sqrt{2}} = 1 - \lambda_{1/\sqrt{2}}, \tag{8.58}$$

where λ_r is given by (8.47). From (8.58), we immediately deduce that

$$\lambda_{\sqrt{2}}(1 - \lambda_{\sqrt{2}}) = (1 - \lambda_{1/\sqrt{2}})\lambda_{1/\sqrt{2}}. \tag{8.59}$$

Next, from (8.48), we deduce the identity

$$\sqrt{2} = \frac{F(1 - \lambda_{\sqrt{2}})}{F(\lambda_{\sqrt{2}})},$$

or

$$F(1 - \lambda_{\sqrt{2}}) = \sqrt{2}F(\lambda_{\sqrt{2}}). \tag{8.60}$$

We now let $r = \sqrt{2}$ in (8.51) and apply (8.60) to deduce that

$$\frac{1}{\pi} = \lambda_{\sqrt{2}}(1 - \lambda_{\sqrt{2}})F(\lambda_{\sqrt{2}})\{F'(1 - \lambda_{\sqrt{2}}) + \sqrt{2}F'(\lambda_{\sqrt{2}})\}. \tag{8.61}$$

Next, substituting $\beta = \lambda_{\sqrt{2}}$ and $\lambda = \lambda_{1/\sqrt{2}}$ into (8.56) and simplifying using (8.58), (8.59) and (8.60), followed by multiplying the resulting identity by $F(\lambda_{\sqrt{2}})$, we deduce that

$$-\frac{F^2(\lambda_{\sqrt{2}})}{\sqrt{\lambda_{\sqrt{2}}}} + \sqrt{2}(1 + \sqrt{\lambda_{\sqrt{2}}})F(\lambda_{\sqrt{2}})F'(1 - \lambda_{\sqrt{2}})$$

$$- \sqrt{2}(1 + \sqrt{\lambda_{\sqrt{2}}})^2 F(\lambda_{\sqrt{2}})F'(\lambda_{\sqrt{2}}) = 0. \tag{8.62}$$

Eliminating $F(\lambda_{\sqrt{2}})F'(1 - \lambda_{\sqrt{2}})$ using (8.62) and (8.61), we find that

$$\frac{2(1 + \sqrt{\lambda_{\sqrt{2}}})}{\pi} = \frac{\sqrt{2}\lambda_{\sqrt{2}}(1 - \lambda_{\sqrt{2}})}{\sqrt{\lambda_{\sqrt{2}}}}F^2(\lambda_{\sqrt{2}})$$

$$+ \{(1 + \sqrt{\lambda_{\sqrt{2}}})^2 + \sqrt{2}(1 + \sqrt{\lambda_{\sqrt{2}}})\}2\lambda_{\sqrt{2}}(1 - \lambda_{\sqrt{2}})F(\lambda_{\sqrt{2}})F'(\lambda_{\sqrt{2}}). \tag{8.63}$$

We will next evaluate $\lambda_{\sqrt{2}}$. From (8.53) and (8.58), we deduce that

$$2F(\lambda_{\sqrt{2}}) = (1 + \sqrt{1 - \lambda_{1/\sqrt{2}}})F(\lambda_{1/\sqrt{2}}) = (1 + \sqrt{\lambda_{\sqrt{2}}})F(1 - \lambda_{\sqrt{2}}). \tag{8.64}$$

Using (8.60) and (8.64), we find that

$$\lambda_{\sqrt{2}} = (\sqrt{2} - 1)^2. \tag{8.65}$$

Next, from (8.39), we have

$$F^2(t) = {}_3F_2\left(\frac{1}{2}, \frac{1}{2}, \frac{1}{2}; 1, 1; 4t(1-t)\right). \tag{8.66}$$

Substituting $t = \lambda_{\sqrt{2}}$ and using (8.65), we find that

$$4\lambda_{\sqrt{2}}(1 - \lambda_{\sqrt{2}}) = 40\sqrt{2} - 56, \tag{8.67}$$

and hence

$$F^2(\lambda_{\sqrt{2}}) = \sum_{k=0}^{\infty} \frac{(\frac{1}{2})_k^3}{(1)_k^3} (40\sqrt{2} - 56)^k. \tag{8.68}$$

Next, differentiating (8.66) with respect to t, substituting $t = \lambda_{\sqrt{2}}$ and using (8.65), we find that

$$2\lambda_{\sqrt{2}}(1 - \lambda_{\sqrt{2}})F'(\lambda_{\sqrt{2}})F(\lambda_{\sqrt{2}}) = (1 - 2\lambda_{\sqrt{2}}) \sum_{k=0}^{\infty} \frac{(\frac{1}{2})_k^3}{(1)_k^3} k(40\sqrt{2} - 56)^k. \tag{8.69}$$

Substituting (8.68) and (8.69) into (8.63), we obtain the following Ramanujan-type series for $1/\pi$:

$$\frac{1}{\pi} = \sum_{k=0}^{\infty} \frac{(\frac{1}{2})_k^3}{(1)_k^3} (40\sqrt{2} - 56)^k ((8 - 5\sqrt{2})k + 3 - 2\sqrt{2}). \tag{8.70}$$

One of the most important facts needed to derive a series similar to (8.70) using the method illustrated in this chapter is an identity relating $\lambda(q)$ and $\lambda(q^p)$ when p is an odd prime. The reader is encouraged to derive the series (due to Ramanujan)

$$\sum_{k=0}^{\infty} (6k+1) \frac{(\frac{1}{2})_k^3}{(1)_k^3} \left(\frac{1}{4}\right)^k = \frac{4}{\pi}. \tag{8.71}$$

Remark 8.6. The series (8.70) is one of many series for $1/\pi$. Such series can be found in a 1859 article by Bauer [8]. Ramanujan, in his famous paper "Modular equations and approximations to π" [62], recorded a total of 17 series similar to (8.70) and their proofs were briefly discussed. For more details of such a series and its history, see [11]. For more examples of such series and their relation with modular forms and functions, see [13] and a recent book by Cooper [40].

Remark 8.7. It can be shown using Theorem 3.2 that

$$4\lambda(q)(1 - \lambda(q)) = \frac{2^6 q}{(-q; q^2)_\infty^{24}}. \tag{8.72}$$

For positive rational number n, the Ramanujan–Weber class invariant is defined by

$$G_n = 2^{-1/4} e^{\pi \sqrt{n}/24} (-e^{\pi \sqrt{n}}; e^{-2\pi \sqrt{n}})_\infty. \tag{8.73}$$

Using (8.72) and (8.73), we may rewrite (8.67) as

$$
\begin{aligned}
G_2^{24} &= \frac{1}{4\lambda_{\sqrt{2}}(1 - \lambda_{\sqrt{2}})} \\
&= \frac{7}{8} + \frac{5}{8}\sqrt{2}.
\end{aligned}
$$

The above proof of (8.70) shows that the values of G_n play an important role in the derivation of Ramanujan's series for $1/\pi$. For more details, see [13, p. 182, (5.5.13)].

Weber [71, pp. 721–726] was the first mathematician to tabulate the values of G_n. Ramanujan, in one of his most influential papers "*Modular equations and approximations to π*" [62], added 31 more evaluations of G_n. In 1998, using class field theory, Kronecker's limit formula and Shimura's reciprocity law, Chan [24] added another 20 values of G_n, the last in the list being G_{2737}. For the relation between class field theory and the Ramanujan–Weber class invariant G_n, see the book by Cox [42, Chapters 2 and 3].

Exercises for Chapter 8

1. (a) Use (1.11) in the form

$$(z - z^{-1})\prod_{k=1}^{\infty}(1 - z^2q^k)(1 - z^{-2}q^k)(1 - q^k) = \sum_{j=-\infty}^{\infty}(-1)^j z^{2j+1}q^{j(j+1)/2}$$

to deduce the following identities:

$$(z - z^{-1})\prod_{k=1}^{\infty}(1 - z^2q^k)(1 - z^{-2}q^k)(1 - q^k)$$

$$= \sum_{j=-\infty}^{\infty} z^{4j+1}q^{2j^2+j} - \sum_{j=-\infty}^{\infty} z^{2j-1}q^{2j^2-j}$$

$$= z\prod_{k=1}^{\infty}(1 + z^4q^{4k-1})(1 + z^{-4}q^{4k-3})(1 - q^{4k})$$

$$- z^{-1}\prod_{k=1}^{\infty}(1 + z^4q^{4k-3})(1 + z^{-4}q^{4k-1})(1 - q^{4k}). \quad (8.74)$$

(b) Use (8.74) to deduce that

$$\prod_{k=1}^{\infty}(1 - q^k)^3 = \prod_{k=1}^{\infty}(1 + q^{4k-3})(1 + q^{4k-1})(1 - q^{4k})$$

$$\times \left(1 - 4\sum_{j=1}^{\infty}\left(\frac{q^{4j-3}}{1 + q^{4j-3}} - \frac{q^{4j-1}}{1 + q^{4j-1}}\right)\right).$$

(c) Deduce that if $r_2(n)$ denotes the number of representations of a positive integer n as a sum of two squares, then

$$r_2(n) = 4\left(\sum_{\substack{d|n \\ d\equiv 1 \ (\text{mod } 4)}} 1 - \sum_{\substack{d|n \\ d\equiv 3 \ (\text{mod } 4)}} 1\right).$$

The above proof of the formula for $r_2(n)$ is due to Hirschhorn [49]. We have seen a proof of Lagrange's four-square theorem in this chapter. Hirschhorn discovered a proof of Lagrange's theorem using a similar technique to the one illustrated in the above exercise. For more details, see Hirschhorn's article [50].

2. (a) Use (2.35) to deduce that

$$((1 - \lambda)(1 - \beta))^{1/4} + (\lambda\beta)^{1/4} = 1, \quad (8.75)$$

where

$$\lambda = \lambda(q) \quad \text{and} \quad \beta = \lambda(q^3).$$

(b) Prove (8.72).

(c) Use (a) and (b) to show that

$$2^{-6}e^{\pi\sqrt{3}}\prod_{k=1}^{\infty}\left(1+e^{-\sqrt{3}\pi(2k-1)}\right)^{24} = 4. \tag{8.76}$$

Hint: You will need (7.30) and (8.73).

3. Use (8.75), (8.76) and the method discussed in this chapter to derive (8.71).
4. Show that (8.43) implies (8.42).

9 The Gauss–Brent–Salamin algorithm for π

In this chapter, we introduce two simple sequences arising from arithmetic and geometric means of two numbers. We show that these two sequences converge to a common limit, known as Gauss' arithmetic–geometric mean, and derive a sequence that converges rapidly to π using these sequences.

9.1 Gauss' arithmetic–geometric mean

Theorem 9.1. *Let $a, b \in \mathbf{R}^+$. Define*

$$a_{n+1} = \frac{a_n + b_n}{2}$$

and

$$b_{n+1} = \sqrt{a_n b_n}, \quad n \geq 0.$$

Then, for $n \geq 1$,

$$b_n \leq b_{n+1} \leq a_{n+1} \leq a_n. \tag{9.1}$$

Proof. We first recall the arithmetic–geometric inequality

$$\frac{a + b}{2} \geq \sqrt{ab}.$$

This yields the second inequality of (9.1), namely,

$$b_{n+1} \leq a_{n+1}.$$

Next,

$$b_{n+1} = \sqrt{a_n b_n} \geq \sqrt{b_n b_n} = b_n,$$

giving the first inequality of (9.1). Finally,

$$a_{n+1} = \frac{a_n + b_n}{2} \leq \frac{a_n + a_n}{2} = a_n$$

and this completes the proof of the theorem. $\qquad \square$

https://doi.org/10.1515/9783110541915-009

Our initial values a, b may not satisfy the inequality $b_0 \leq a_0$. However, for $n \geq 1$, (9.1) holds. From (9.1), we find that $\{a_n | n \in \mathbf{Z}^+\}$ is a monotone decreasing sequence in $\mathbf{R}$ bounded below by b_1. This implies that $A = \lim_{n \to \infty} a_n$ exists by the monotone convergence theorem. Similarly, $\{b_n | n \in \mathbf{Z}^+\}$ is a monotone increasing sequence in $\mathbf{R}$ bounded above by a_1 and $B = \lim_{n \to \infty} b_n$ exists. Since

$$a_{n+1} = \frac{a_n + b_n}{2},$$

by letting $n \to \infty$, we deduce that

$$A = \frac{A + B}{2}$$

and this implies that $A = B$. We denote this common limit of the two sequences $\{a_n | n \in \mathbf{Z}^+\}$ and $\{b_n | n \in \mathbf{Z}^+\}$ by $M(a, b)$ since the limit depends on the initial values a and b of the sequences. Note that $M(a, b)$ can also be written as $M(a_1, b_1)$, since the sequences defined by $a'_n = a_{n-1}$ and $b'_n = b_{n-1}$ has the same limits A and B as $n \to \infty$. Therefore, we may assume in the beginning that $b \leq a$.

The function $M(a, b)$ is known as Gauss' arithmetic–geometric mean. The following are two important properties of $M(a, b)$.

Theorem 9.2.
(a) *If $a, b, \lambda \in \mathbf{R}^+$ with $b \leq a$, then*

$$M(\lambda a, \lambda b) = \lambda M(a, b). \tag{9.2}$$

(b) *If $a, b \in \mathbf{R}^+$ with $b \leq a$, then*

$$M(a, b) = \left(\frac{a + b}{2} \right) M\left(1, \frac{2\sqrt{ab}}{a + b} \right). \tag{9.3}$$

In particular, if $x \leq 1$,

$$M(1, x) = \left(\frac{1 + x}{2} \right) M\left(1, \frac{2\sqrt{x}}{1 + x} \right). \tag{9.4}$$

Proof. To prove (a), we note that if we define another sequence $\{a_n^*\}$ and $\{b_n^*\}$ with initial values λa and λb, then, by induction, we observe that

$$a_n^* = \lambda a_n \quad \text{and} \quad b_n^* = \lambda b_n.$$

Hence,

$$M(\lambda a, \lambda b) = \lim_{n \to \infty} a_n^* = \lim_{n \to \infty} \lambda a_n = \lambda M(a, b).$$

To prove (b), we set $a_n^\dagger = a_{n+1}$ and $b_n^\dagger = b_{n+1}$. This implies that

$$M(a_0^\dagger, b_0^\dagger) = M\left(\frac{a+b}{2}, \sqrt{ab}\right).$$

But

$$\lim_{n\to\infty} a_n^\dagger = \lim_{n\to\infty} a_n$$

and hence

$$M(a, b) = M\left(\frac{a+b}{2}, \sqrt{ab}\right).$$

By (9.2), we find that

$$M\left(\frac{a+b}{2}, \sqrt{ab}\right) = \left(\frac{a+b}{2}\right) M\left(1, \frac{2\sqrt{ab}}{a+b}\right)$$

and this concludes the proof of (9.3).

Finally, by setting $a = 1$ and $b = x$ in (9.3), we deduce (9.4). $\qquad\square$

9.2 Gauss' AGM, Jacobi's theta functions and hypergeometric series

Jacobi's theta functions offer a natural example of the sequences $\{a_n\}$ and $\{b_n\}$. For $n \geq 0$, let

$$a_n = \Theta_3^2(e^{-2^n \pi v}) \quad \text{and} \quad b_n = \Theta_4^2(e^{-2^n \pi v}).$$

Then, from (2.1) and (2.10), we observe that, for $n \geq 1$,

$$a_{n+1} = \frac{a_n + b_n}{2} \quad \text{and} \quad b_{n+1} = \sqrt{a_n b_n}.$$

Observe that

$$M(a_0, b_0) = M(a_1, b_1) = \cdots = M(a_n, b_n).$$

By (9.2), we find that

$$\Theta_3^2(e^{-\pi v}) M\left(1, \frac{\Theta_4^2(e^{-\pi v})}{\Theta_3^2(e^{-\pi v})}\right) = \Theta_3^2(e^{-2^n \pi v}) M\left(1, \frac{\Theta_4^2(e^{-2^n \pi v})}{\Theta_3^2(e^{-2^n \pi v})}\right).$$

Now,

$$\lim_{n \to \infty} \Theta_3^2(e^{-2^n \pi \nu}) = 1$$

and

$$\lim_{n \to \infty} \Theta_4^2(e^{-2^n \pi \nu}) = 1.$$

Hence,

$$\Theta_3^2(e^{-\pi \nu}) M\left(1, \frac{\Theta_4^2(e^{-\pi \nu})}{\Theta_3^2(e^{-\pi \nu})}\right) = M(1,1) = 1.$$

In other words,

$$\Theta_3^2(e^{-\pi \nu}) = \frac{1}{M\left(1, \dfrac{\Theta_4^2(e^{-\pi \nu})}{\Theta_3^2(e^{-\pi \nu})}\right)}. \tag{9.5}$$

We now recall from (2.30) and (2.23) that

$$\frac{\Theta_4^2(q)}{\Theta_3^2(q)} = \sqrt{1 - \lambda(q)}.$$

Consequently, we can rewrite (9.5) as

$$\Theta_3^2(e^{-\pi \nu}) = \frac{1}{M(1, \sqrt{1 - \lambda(e^{-\pi \nu})})}.$$

Next, from (8.29), we deduce that

$$_2F_1\left(\frac{1}{2}, \frac{1}{2}; 1; \lambda(e^{-\pi \nu})\right) = \frac{1}{M(1, \sqrt{1 - \lambda(e^{-\pi \nu})})}. \tag{9.6}$$

Using (9.4) and (9.6), we deduce that

$$_2F_1\left(\frac{1}{2}, \frac{1}{2}; 1; \lambda\right) = \frac{1}{M(1, \sqrt{1 - \lambda})}$$

$$= \frac{1}{M\left(\dfrac{1 + \sqrt{1 - \lambda}}{2}, (1 - \lambda)^{1/4}\right)}$$

$$= \frac{2}{1 + \sqrt{1 - \lambda}} \frac{1}{M\left(1, \dfrac{2(1 - \lambda)^{1/4}}{1 + \sqrt{1 - \lambda}}\right)}$$

$$= \frac{2}{1 + \sqrt{1 - \lambda}} \, {}_2F_1\left(\frac{1}{2}, \frac{1}{2}; 1; y\right)$$

where

$$\sqrt{1 - y} = \frac{2(1 - \lambda)^{1/4}}{1 + \sqrt{1 - \lambda}}$$

or

$$y = \left(\frac{1 - \sqrt{1 - \lambda}}{1 + \sqrt{1 - \lambda}}\right)^2.$$

In other words, we find that

$$_2F_1\left(\frac{1}{2}, \frac{1}{2}; 1; \lambda\right) = \frac{2}{1 + \sqrt{1 - \lambda}} \, {}_2F_1\left(\frac{1}{2}, \frac{1}{2}; 1; \left(\frac{1 - \sqrt{1 - \lambda}}{1 + \sqrt{1 - \lambda}}\right)^2\right). \tag{9.7}$$

Since (9.7) is valid for $0 < \lambda < 1$, we may replace λ by t and deduce the following corollary.

Corollary 9.3. *For all real positive numbers t sufficiently near 0, we have*

$$_2F_1\left(\frac{1}{2}, \frac{1}{2}; 1; t\right) = \frac{2}{1 + \sqrt{1 - t}} \, {}_2F_1\left(\frac{1}{2}, \frac{1}{2}; 1; \left(\frac{1 - \sqrt{1 - t}}{1 + \sqrt{1 - t}}\right)^2\right). \tag{9.8}$$

If we let

$$x = \frac{1 - \sqrt{1 - t}}{1 + \sqrt{1 - t}},$$

then we obtain from (9.8) the identity

$$_2F_1\left(\frac{1}{2}, \frac{1}{2}; 1; 1 - \left(\frac{1 - x}{1 + x}\right)^2\right) = (1 + x) \, 2F_1\left(\frac{1}{2}, \frac{1}{2}; 1; x^2\right). \tag{9.9}$$

Remark 9.1. The transformation formula (9.9) has the following generalization [3, p. 128]:

$$_2F_1\left(a, b; 2b; 1 - \left(\frac{1 - x}{1 + x}\right)^2\right) = (1 + x)^{2a} \, {}_2F_1\left(a, a + \frac{1}{2} - b; a + \frac{1}{2}; x^2\right). \tag{9.10}$$

The proof of (9.10) is entirely different from the approach (which only applies to the case $a = 1/2$) illustrated in this section.

9.3 The Gauss–Brent–Salamin algorithm for π

We now state the Gauss–Brent–Salamin algorithm for π.

Theorem 9.4. *Let $a_0 = 1$ and $b_0 = 1/\sqrt{2}$. For $n \geq 0$, let*

$$a_{n+1} = \frac{a_n + b_n}{2},$$
$$b_{n+1} = \sqrt{a_n b_n},$$

and

$$\pi_n = \frac{2a_n^2}{1 - \sum_{j=0}^{n} 2^j (a_j^2 - b_j^2)}.$$

Then π_n approaches π as n approaches ∞.

The first eight iterations of the above algorithm produce 1, 3, 8, 19, 41, 84, 172 and 346 digits of π. For a discussion of the rate of convergence of this sequence, see [13, p. 48].

Theorem 9.4 was discovered independently by Brent [18] and Salamin [65] around 1976. A generalization of Theorem 9.4 can be found in the work of Gauss. For more details of the fascinating discovery of the origin of Theorem 9.4 and Gauss' generalization, see [6, Chapter 7]. Recently, new generalizations of Theorem 9.4 have been found. For more details, see the article by Chan [28, p. 89].

The original proof of Theorem 9.4 uses elliptic integrals. In this section, we will derive a proof of Theorem 9.4 using results that we have obtained in the previous chapters. Our approach is a modification of the proof of Brent and Salamin presented in [13] but avoids the use of elliptic integrals.

Proof of Theorem 9.4. We need to record a few identities. Observe from (7.30),

$$\lambda(e^{-\pi}) = 1/2. \tag{9.11}$$

Substituting this value into (8.51), we deduce that

$$F(1/2)F'(1/2) = \frac{2}{\pi}. \tag{9.12}$$

Next, set

$$G(\lambda_r) = 2\lambda_r(1 - \lambda_r)F'(\lambda_r) + (1 - \lambda_r)F(\lambda_r), \tag{9.13}$$

where λ_r is defined as in (8.47). From (9.13), we find that

$$2\lambda_r(1 - \lambda_r)F'(\lambda_r) = G(\lambda_r) - (1 - \lambda_r)F(\lambda_r). \tag{9.14}$$

Substituting (9.14) into (8.56) and simplifying using (2.32) and (8.53), we conclude that

$$G(\lambda_r) = \left(1 + \sqrt{1 - \lambda_r}\right)G(\beta_r) - \sqrt{1 - \lambda_r}F(\lambda_r), \tag{9.15}$$

where $\beta_r = \lambda(e^{-2r\pi})$. As in the previous section, we note that if

$$A_n = \Theta_3^2(e^{-2^n\pi}) \tag{9.16}$$

and

$$B_n = \Theta_4^2(e^{-2^n\pi}), \tag{9.17}$$

then, by (2.1) and (2.10),

$$A_{n+1} = \frac{A_n + B_n}{2}, \quad \text{and} \quad B_{n+1} = \sqrt{A_n B_n}.$$

Let

$$\alpha_n = \left(\frac{C_n}{A_n}\right)^2,$$

where

$$C_n = \Theta_2^2(e^{-2^n\pi}).$$

Observe that, by (8.29),

$$A_n = F(\lambda(e^{-2^n\pi})) \tag{9.18}$$

and, by (2.30),

$$\alpha_n = \lambda(e^{-2^n\pi}). \tag{9.19}$$

By (2.23), (9.16) and (9.17), we have

$$\sqrt{1 - \alpha_n} = \frac{B_n}{A_n}. \tag{9.20}$$

Next, by letting $r = 2^n$ in (9.15) and using (9.20), we deduce that

$$G(\alpha_n) = \left(1 + \frac{B_n}{A_n}\right)G(\alpha_{n+1}) - \frac{B_n}{A_n}F(\alpha_n)$$

or

$$A_n G(\alpha_n) = 2A_{n+1}G(\alpha_{n+1}) - B_n A_n \frac{F(\alpha_n)}{A_n}.$$

Now, (8.29) and (9.16) show that, for all $n \geq 0$,

$$\frac{F(\alpha_n)}{A_n} = 1.$$

Therefore,

$$2A_{n+1}G(\alpha_{n+1}) - A_n G(\alpha_n) = A_n B_n. \tag{9.21}$$

Using the elementary identity

$$ab = 2\left(\frac{a+b}{2}\right)^2 - a^2 + \frac{a^2 - b^2}{2},$$

we find that

$$A_n B_n = 2A_{n+1}^2 - A_n^2 - \frac{A_n^2 - B_n^2}{2}. \tag{9.22}$$

Combining (9.21) and (9.22) and rearranging, we deduce that

$$2(A_{n+1}G(\alpha_{n+1}) - A_{n+1}^2) - (A_n G(\alpha_n) - A_n^2) = 2^{-1}(A_n^2 - B_n^2). \tag{9.23}$$

Multiplying (9.23) by 2^n, we find that

$$2^{n+1}(A_{n+1}G(\alpha_{n+1}) - A_{n+1}^2) - 2^n(A_n G(\alpha_n) - A_n^2) = 2^{n-1}(A_n^2 - B_n^2). \tag{9.24}$$

We next sum both sides of (9.24) from 0 to N to deduce that

$$2^{N+1}(A_{N+1}G(\alpha_{N+1}) - A_{N+1}^2) - (A_0 G(\alpha_0) - A_0^2) = \sum_{n=0}^{N} 2^{n-1}(A_n^2 - B_n^2). \tag{9.25}$$

Now, from (9.13) and the fact that

$$\alpha_n = e^{-2^n \pi} + \cdots,$$

we find that

$$\lim_{N \to \infty} 2^{N+1}(A_{N+1}G(\alpha_{N+1}) - A_{N+1}^2) = 0.$$

Therefore, letting N tend to infinity in (9.25), we arrive at

$$-A_0 G(\alpha_0) + A_0^2 = \sum_{n=0}^{\infty} 2^{n-1}(A_n^2 - B_n^2)$$

or

$$A_0 G(\alpha_0) = A_0^2 \left(1 - \sum_{n=0}^{\infty} 2^{n-1}((A_n/A_0)^2 - (B_n/A_0)^2) \right). \tag{9.26}$$

By (9.11), (9.18) and (9.19), we conclude that

$$\alpha_0 = \lambda(e^{-\pi}) = 1/2$$

and

$$A_0 = F(1/2). \tag{9.27}$$

Substituting the last two identities into (9.26), we deduce that

$$F(1/2)\left(\frac{1}{2}F'(1/2) + \frac{1}{2}F(1/2) \right)$$

$$= F^2(1/2)\left(1 - \sum_{n=0}^{\infty} 2^{n-1}((A_n/A_0)^2 - (B_n/A_0)^2) \right).$$

Using (9.12) and (9.27), we deduce that

$$\frac{2}{\pi} = F^2(1/2)\left(1 - \sum_{n=0}^{\infty} 2^n((A_n/A_0)^2 - (B_n/A_0)^2) \right). \tag{9.28}$$

Let

$$a_n = A_n/A_0 \quad \text{and} \quad b_n = B_n/A_0. \tag{9.29}$$

Then, from (9.16) and (8.29), we find that

$$F(\alpha_n)A_0 = F(\alpha_0)A_n.$$

From the theta function representation of $F(\alpha_n)$ (see (8.29)), we conclude that

$$\lim_{n \to \infty} F(\alpha_n) = 1.$$

Hence,

$$\lim_{n \to \infty} a_n = \lim_{n \to \infty} \frac{A_n}{A_0} = \frac{1}{F(1/2)}. \tag{9.30}$$

From (9.28)–(9.30), we deduce that

$$\frac{1}{\pi} = \lim_{N \to \infty} \frac{1}{2a_N^2} \left(1 - \sum_{j=0}^{N} 2^j (a_j^2 - b_j^2) \right)$$

and the proof of Theorem 9.4 is complete. □

Remark 9.2. The definition of $G(\lambda)$ in (9.13) is motivated by the identity found in Borweins' book [13, Theorem 1.3(b)]. It can be replaced by

$$G_{s,t}(\lambda_r) = s\lambda_r(1 - \lambda_r)F'(\lambda_r) + t(1 - \lambda_r)F(\lambda_r).$$

For each choice of s and t, there is an expression similar to (9.15) and with probably some exceptions, an algorithm similar to Theorem 9.4.

The choice of s and t determines the simplicity of the analogue of (9.15) and it turns out that $s = 2$ and $t = 1$ yields a relatively simple expression (9.15) and the resulting algorithm coincides with Theorem 9.4.

We conclude this chapter with Borweins' algorithm (with some modifications) [13, p. 170] for computing π from Theorem 9.4.

Theorem 9.5. *Let $t_0 = 1/4$ and $s_0 = 1/2$,*

$$s_n = \left(\frac{1 - \sqrt{1 - s_{n-1}}}{1 + \sqrt{1 - s_{n-1}}} \right)^2$$

and

$$t_n = \left(\frac{2}{1 + \sqrt{1 - s_{n-1}}} \right)^2 t_{n-1} - 2^{n-1} s_n.$$

Then t_n^{-1} converges to π.

Proof. Let

$$t_n = \frac{1}{\pi_n} = \frac{1 - \sum_{j=0}^{n} 2^j (a_j^2 - b_j^2)}{2a_n^2}.$$

Let

$$s_n = \alpha(e^{-2^n \pi}).$$

Note that $t_0 = 1/4$ and $s_0 = 1/2$. Furthermore,

$$2a_{n+1}^2 t_{n+1} - 2a_n^2 t_n = -2^{n+1}(a_{n+1}^2 - b_{n+1}^2).$$

This implies that

$$t_{n+1} = \frac{a_n^2}{a_{n+1}^2} - 2^n\left(1 - \frac{b_{n+1}^2}{a_{n+1}^2}\right). \tag{9.31}$$

By (2.2), we observe that

$$\frac{a_n^2}{a_{n+1}^2} = \left(\frac{2}{1 + \sqrt{1 - s_n}}\right)^2. \tag{9.32}$$

Moreover,

$$1 - \frac{b_{n+1}^2}{a_{n+1}^2} = s_{n+1}. \tag{9.33}$$

Substituting (9.32) and (9.33) into (9.31), we complete the proof of Theorem 9.5. $\square$

The first eight iterations of the above algorithm produce 1, 4, 8, 19, 42, 84, 173 and 346 digits of π. For a discussion of the rate of convergence of this sequence, see [13, pp. 169–170].

Remark 9.3. For every algorithm similar to Theorem 9.4, we can derive an analogue of Theorem 9.5. However, not all analogues of Theorem 9.5 follow from analogues of Theorem 9.4. In fact, Borwein and Garvan [16] and Chan [25] discovered many algorithms similar to Theorem 9.5 which do not follow from analogues of Theorem 9.4. An example of such iterations which is similar to Theorem 9.5 is the following iteration given by Chan [25, Iteration 1.5].

Theorem 9.6. *Let $t_0 = 0$ and $s_0 = 1/\sqrt{2}$,*

$$s_n = \frac{1 - \sqrt{1 - s_{n-1}^2}}{1 + \sqrt{1 - s_{n-1}^2}}$$

and

$$t_n = (1 + s_n)^2 t_{n-1} + 2^n(1 - s_n)s_n.$$

Then t_n^{-1} converges to π.

Having two different algorithms that converge rapidly to π allows us to double check the digits we obtained from our computations.

It appears that analogues of Theorem 9.4 are rarer than those of Theorem 9.5. One reason is perhaps that algorithms of Gauss–Brent–Salamin type require three modular forms of weight 1 (such as $\Theta_j^2(q)$, $j = 2, 3, 4$), while algorithms of the second type require a modular function (such as $\lambda(q)$) and a modular form of weight 1 (such as $\Theta_3^2(q)$).

Exercises for Chapter 9

1. Let $a, b \in \mathbf{R}^+$. Let

$$T(a,b) = \frac{2}{\pi} \int_0^{\pi/2} \frac{d\theta}{\sqrt{a^2 \cos^2 \theta + b^2 \sin^2 \theta}}.$$

(a) Show that

$$T(a,b) = \frac{1}{\pi} \int_0^\infty \frac{dt}{\sqrt{(a^2 + t^2)(b^2 + t^2)}}.$$

Hint: Use the substitution $t = b \tan \theta$.

(b) Use the substitution $u = \frac{1}{2}(t - ab/t)$ to show that

$$T(a,b) = T((a+b)/2, \sqrt{ab}). \tag{9.34}$$

Remark: One should pay attention to the lower and upper limits of the integrals before using the given substitution.

(c) Let a_n and b_n be the sequences defined in Theorem 9.1, with $a_1 = a$ and $b_1 = b$. Let $M(a,b)$ be the common limit of a_n and b_n as $n \to \infty$. Use (9.34) to show that

$$T(a,b) = \frac{1}{M(a,b)}. \tag{9.35}$$

Remark: When $a = 1$ and $b = x$, we may rewrite (9.35), using

$${}_2F_1(1/2, 1/2; 1; k^2) = \frac{2}{\pi} \int_0^{\pi/2} \frac{d\theta}{\sqrt{1 - k^2 \sin^2 \theta}},$$

as

$${}_2F_1(1/2, 1/2; 1; k^2) = \frac{1}{M(1, \sqrt{1 - k^2})}.$$

This last identity is equivalent to (9.6).

2. Let

$$E(x) = {}_2F_1(-1/2, 1/2; 1; x) \quad \text{and} \quad K(x) = {}_2F_1(1/2; 1/2; 1; x).$$

(a) Verify that

$$E(x) = \frac{2}{\pi} \int_0^{\pi/2} \sqrt{1 - x \sin^2 \theta} \, d\theta$$

and

$$K(x) = \frac{2}{\pi} \int_0^{\pi/2} \frac{d\theta}{\sqrt{1 - x\sin^2\theta}}$$

and use the integral representations of $E(x)$ and $K(x)$ to show that

$$x\frac{dE(x)}{dx} = \frac{E(x) - K(x)}{2}. \tag{9.36}$$

(b) Let

$$D_x = x\frac{d}{dx}.$$

Show that

$$(D_x^2 - x(D_x - 1/2)(D_x + 1/2))E(x) = 0 \tag{9.37}$$

and verify that

$$[D_x + 1/2 - x(D_x + 1/2)](D_x - 1/2)E(x) = -\frac{E(x)}{4}. \tag{9.38}$$

Remark: The differential equation (9.37) is similar to (8.25). In general, if

$$y = {}_2F_1(a, b; c; x),$$

then

$$[D_x(D_x + c - 1) - x(D_x + a)(D_x + b)]y = 0.$$

(c) Use (9.36) and (9.38) to show that

$$x\frac{dK(x)}{dx} = \frac{E(x) - (1 - x)K(x)}{2(1 - x)}. \tag{9.39}$$

Remark: Identities (9.36) and (9.39) play important roles in the original proof of Theorem 9.4. For more details, see [13, Section 1.4]. This problem is motivated by the approach to (9.39) given in [3, p. 96].

3. Fill in the details for the derivation of (9.15).

4. Derive the iteration given in Theorem 9.6.

Index

Bibliography

[1]   G. L. Alexanderson and G. Pólya, *Gaußian binomial coefficients*, Elem. Math., **26** (1971), 102–109.

[2]   G. E. Andrews, *A simple proof of Jacobi's triple product identity*, Proc. Amer. Math. Soc., **16** (1965), 333–334.

[3]   G. E. Andrews, R. Askey and R. Roy, *Special Functions*, Encyclopedia of Mathematics and Its Applications, The University Press, Cambridge, 1999.

[4]   S. Ahlgren and M. Boylan, *Arithmetic properties of the partition function*, Inventiones Mathematicae, **153** (2003), no. 3, 487–502.

[5]   T. M. Apostol, *Modular Functions and Dirichlet Series in Number Theory*, 2nd edition, Springer-Verlag, New York, 1990.

[6]   J. Arndt and C. Haenel, *π-Unleashed*, Springer-Verlag, Germany, 2001.

[7]   J. Bak and D. J. Newman, *Complex Analysis*, 2nd edition, Springer-Verlag, New York, 1997.

[8]   G. Bauer, *Von den Coefficienten der Reihen von Kugelfunctionen einer Variabeln*, J. Reine Angew. Math., **56** (1859), 101–121.

[9]   L. Berggren, J. M. Borwein and P. B. Borwein, *Pi: A Source Book*, Springer-Verlag, New York, 1997.

[10] B. C. Berndt, *Ramanujan's Notebooks Part III*, Springer-Verlag, New York, 1991.

[11] B. C. Berndt, N. D. Baruah and H. H. Chan, *Ramanujan's series for 1/π: A survey*, Math. Student (2008), 1–24. Special Centenary Volume (2007); and Amer. Math. Monthly, **116** (2009), no. 7, 567–587.

[12] B. C. Berndt and A. J. Yee, *A page on Eisenstein series in Ramanujan's lost notebook*, Glasgow Math. J., **45** (2003), 123–129.

[13] J. M. Borwein and P. B. Borwein, *Pi and the AGM; A Study in Analytic Number Theory and Computational Complexity*, Wiley, New York, 1987.

[14] J. M. Borwein and P. B. Borwein, *A cubic counterpart of Jacobi's identity and the AGM*, Trans. Amer. Math. Soc., **323** (1991), 691–701.

[15] J. M. Borwein, P. B. Borwein and F. G. Garvan, *Some cubic modular identities of Ramanujan*, Trans. Amer. Math. Soc., **343** (1994), 35–47.

[16] J. M. Borwein and F. G. Garvan, *Approximations to π via the Dedekind eta function*, In: CMS Conf. Proc., **20**, Amer. Math. Soc., Providence, RI, 1997, 89–115.

[17] F. Brafman, *Generating functions of Jacobi and related polynomials*, Proc. Amer. Math. Soc., **2** (1951), 942–949.

[18] R. P. Brent, *Fast multiple-precision evaluation of elementary functions*, J. ACM, **23** (1976), 242–251.

[19] T. J. I'A. Bromwich, *An Introduction to the Theory of Infinite Series*, 2nd revised edition, Macmillian, London, 1964.

[20] L. Carlitz and M. V. Subbarao, *A simple proof of the quintuple product identity*, Proc. Amer. Math. Soc., **32** (1972), no. 1, 42–44.

[21] R. Carter, *Lie Algebras of Finite and Affine Type*, Cambridge University Press, London, 2005.

[22] A. Cauchy, *Oeuvres, ser. 1, vol. 8*, Gauthier-Villars, Paris, 1893.

[23] H. C. Chan, *An Invitation to q-Series: From Jacobi's Triple Product Identity to Ramanujan's "Most Beautiful Identity"*, World Scientific Publishing Company, 2011.

[24] H. H. Chan, *Ramanujan's class invariants and Watson's empirical process*, Journal of

https://doi.org/10.1515/9783110541915-010

London Mathematical Society, Ser. 2, **57**, (1998), 545–561.

[25] H. H. Chan, *Ramanujan's elliptic functions to alternative bases and approximations to π*, In: B. Berndt, et al. (eds.) Number Theory for the Millennium I (Proceedings of the Millennial Number Theory Conference) (Champaign-Urbana, Illinois, May 2000), A. K. Peters Ltd, 2002.

[26] H. H. Chan, *Triple product identity, quintuple product identity and Ramanujan's differential equations for the classical Eisenstein series*, Proc. of the Amer. Math. Soc., **135**, (2007), no. 7, 1987–1992.

[27] H. H. Chan, *The Bailey-Brafman identity and its analogues*, Journal of Mathematical Analysis and Applications, **399** (2013), 12–16.

[28] H. H. Chan, *Analogues of the Brent-Salamin algorithm for evaluating π*, Ramanujan J., **38** (2015), no. 1, 75–100.

[29] H. H. Chan, S. H. Chan and S. Cooper, *The q-binomial theorem*, Math. Medley, **33** (2006), no. 2, 2–6.

[30] H. H. Chan, S. Cooper and P. C. Toh, *Ramanujan's Eisenstein series and powers of Dedekind's eta-function*, J. Lond. Math. Soc. (2), **75** (2007), no. 1, 225–242.

[31] H. H. Chan and R. P. Lewis, *Partition identities and Congruences associated with the Fourier coefficients of the Euler products*, Journal of Computational and Applied Math., **160** (2003), 69–75.

[32] H. H. Chan and C. Krattenthaler, *Recent progress in the study of the representations of integers as sums of squares*, Bull. of the London Math. Soc., **37** (2005), 818–826.

[33] H. H. Chan and Y. Tanigawa, *A generalization of a Brafman-Bailey type identity*, Proc. Amer. Math. Soc., **143**, (2015), no. 1, 185–195.

[34] H. H. Chan, Y. Tanigawa, Y. F. Yang and W. Zudilin, *New analogues of Clausen's identities arising from the theory of modular forms*, Adv. Math., **228** (2011), no. 2, 1294–1314.

[35] H. H. Chan, J. Wan and W. Zudilin, *Legendre polynomials and Ramanujan-type series for 1/π*, Israel J. Math., **194** (2013), no. 1, 183–207.

[36] H. H. Chan and L. Q. Wang, *Borweins' cubic theta functions revisited*, preprint.

[37] H. H. Chan, L. Q. Wang and Y. F. Yang, *Modular forms and k-colored generalized Frobenius partitions*, Trans. Amer. Math. Soc., **371** (2019), no. 3, 2159–2205.

[38] S.-J. Chen and W. Q. Wang, *Dualities and Representations of Lie Superalgebras*, Graduate Studies in Mathematics, **144**, Amer. Math. Soc., Providence, 2012.

[39] S. Cooper, *The quintuple product identity*, Int. J. Number Theory, **2** (2006), 115–161.

[40] S. Cooper, *Ramanujan's Theta Functions*, Springer International Publishing, 2017.

[41] E. T. Copson. *Theory of Functions of a Complex Variable*, Oxford University Press, 1948.

[42] D. A. Cox, *Primes of the Form $x^2 + ny^2$*, 2nd edition, Wiley, New York, 2013.

[43] D. Eichhorn, and K. Ono, *Congruences for partition functions*, In: Analytic Number Theory, Vol. 1 (Allerton Park, IL, 1995), Progr. Math., **138**, Birkhauser Boston, Boston, MA, 1996, 309–321.

[44] G. Gasper and M. Rahman, *Basic Hypergeometric Series*, Cambridge University Press, Cambridge, 1990.

[45] E. Grosswald, *Representations of Integers as Sums of Squares*, Springer-Verlag, New York, 1985.

[46] G.-H. Halphen, *Traité des Fonctions Elliptiques et de leurs Applications, Part 1*, Gauthier-Villars, Paris, 1886.

[47] G. H. Hardy and S. Ramanujan, *Asymptotic formulae in combinatory analysis*, Proc. Lond. Math. Soc., **17** (1918), 75–115.